Differential Display

The Practical Approach Series

Related **Practical Approach** Series Titles

Please see the **Practical Approach** series website at
http://www.oup.co.uk/pas
for full contents lists of all Practical Approach titles.

Differential Display

A Practical Approach

Edited by

R. A. Leslie

Smith Kline Beecham Pharmaceuticals
New Frontiers Science Park (North)
Third Avenue
Harlow
Essex CM19 5AW

and

H. A. Robertson

Department of Pharmacology
Sir Charles Tupper Medical Building
Dalhousie University
Halifax, Nova Scotia
Canada B3H 4H7

OXFORD
UNIVERSITY PRESS

OXFORD
UNIVERSITY PRESS

Great Clarendon Street, Oxford OX2 6DP

Oxford University Press is a department of the University of Oxford.
It furthers the University's objective of excellence in research,
scholarship, and education by publishing worldwide in

Oxford New York

Athens Auckland Bangkok Bogotá Buenos Aires Calcutta Cape Town
Chennai Dar es Salaam Delhi Florence Hong Kong Istanbul Karachi
Kuala Lumpur Madrid Melbourne Mexico City Mumbai Nairobi Paris
São Paulo Singapore Taipei Tokyo Toronto Warsaw

with associated companies in Berlin Ibadan

Oxford is a registered trade mark of Oxford University Press in the UK
and in certain other countries

Published in the United States by Oxford University Press Inc., New York

A catalogue record for this title is available from the British Library

Library of Congress Cataloguing in Publication Data
(Data available)

ISBN 0-19-963759-8 (Hbk.)
ISBN 0-19-963758-x (Pbk.)

1 3 5 7 9 10 8 6 4 2

Typeset in Swift by Footnote Graphics, Warminster, Wilts
Printed in Great Britain on acid-free paper
by The Bath Press, Avon

Preface

When the technique of differential display was first described in the early 1990s, sequence information existed for a few thousand human genes and some functional information was available for most of those. At about that time, the development of high throughput sequencing technologies and large genome databases, including the development of EST (expressed sequence tag) databases and the massive human genome project, greatly accelerated the process of accumulation of genetic sequence data. While amounts of sequence information have been growing logarithmically, however, information about the function of genes has emerged much more slowly. Latest estimates suggest that there are as many as 140,000 human genes in the genome, but to date we have some understanding of the functions of only about 5000 of these. In a very few years the sequences of all human genes will be known, but there will be an information bottleneck regarding the physiological activity of their protein products. Differential gene expression technologies, like differential display, will play an essential role in elucidating these functions.

Differential display was the first differential gene expression screening technology to become accessible to a broad range of academic biologists. As a novel technique, it suffered from some 'teething' problems; for example some laboratories had difficulty reproducing their early results while others found a high rate of false positives. As a result its reputation has suffered, perhaps because in some cases it appeared to be too simple and straightforward so that some early projects were embarked upon without enough preparation. Thus some early experimental paradigms may not have been the most appropriate ones to allow for correct interpretation of the data.

Despite early problems and disappointments, several thousand papers using differential display have now been published. Many improvements in the original technology have been described and more improvements are appearing all the time. This book provides an overview of current differential display technologies and suggests some new ways in which differential gene expression will be studied in the future. Detailed protocols for techniques that have proved successful in practice are supplied. The coverage includes recent advances in fluorescent differential display, experimental design for analysis of complex tissues, the RAP-array technique, efficient display of 3′-end cDNA fragments, and cloning of differentially expressed genes.

Differential display, of course, has not developed in a vacuum and a number of other techniques have been developed to achieve similar aims. For example, techniques such as SAGE or DNA gridding ('microarray') technologies are expected to replace differential display in due course. These technologies are relatively immature however, and equipment is not yet generally available, especially to academic laboratories, for them. Nonetheless, it is important to be aware of these new techniques and their capabilities, and this book descibes some of the new developments in the area.

Ultimately the goal of molecular genetics as a discipline is to unravel the function of all the genes in the human genome (as well as other important genomes). This book will allow researchers to make intelligent decisions about which differential gene expression technique to adopt for their particular project. Step-by-step procedures are given to help ensure that projects use appropriate experimental design, and that the best available techniques are followed in ways most likely to result in success.

Contents

Protocol list

Abbreviations

AP	anchored primers
ARP	arbitrary primers
DD	differential display
DDT	dithiothreitol
DEPC	diethyl pyrocarbonate
DIG	digoxygenin
dNTP	deoxyribonucleoside triphosphates
EST	expressed sequence tag
FA	Fly-Adapter
fluoroDD	fluorescent differential display
GFAP	glial fibrillary acidic protein
IAP	integrin-associated protein
LTP	long-term potentiation
MMP	matrix metalloproteinase
6-OHDA	6-hydroxydopamine
PAGE	polyacrylamide gel electrophoresis
PBS	phosphate-buffered saline
PCR	polymerase chain reaction
PMSF	phenylmethylsulfonyl fluoride
PNK	polynucleotide kinase
RACE	rapid amplification of cDNA ends
RAP-PCR	RNA arbitrarily primed-PCR
RDA	representational difference analysis
RPA	ribonuclease protection assay
RT	reverse transcription
RT-PCR	reverse transcriptase-PCR
SAGE	serial analysis of gene expression
SD	subtractive display
SDD	subtracted differential display
SDS	sodium dodecyl sulfate
SDS–PAGE	sodium dodecyl sulfate–polyacrylamide gel electrophoresis
SH	subtractive hybridization
SSCP	single strand conformation polymorphism

SSH	suppression subtractive hybridization
TEMED	*N,N,N′,N′*-tetramethylethylenediamine
TIMP-1	tissue inhibitor of matrix metalloproteinase-1
TMR	tetramethylrhodamine
UTR	untranslated region
VIP	vasointestinal polypeptide

Chapter 1

An introduction to differential display and related techniques

HAROLD A. ROBERTSON* and RONALD A. LESLIE[†]

*Department of Pharmacology, Dalhousie University, Sir Charles Tupper Medical Building, Halifax, Nova Scotia, B3H 4H7, Canada

[†]Smithkline Beecham Pharmaceuticals, New Frontiers Science Park North, Third Avenue, Harlow, Essex CM19 5AW, UK.

It is worth acknowledging, as there is no way around it, that differential display has been, and remains, a controversial technique. Any technique that engenders articles with the word 'dismay' in the title (1) must be controversial! More than a thousand articles, however, have now been published describing results achieved with the use of the differential display technique and hundreds of genes have been 'discovered'. This book has been written with the firm belief that much of the scepticism that differential display has engendered is based upon a frequent poor understanding of the technique and the premature publication of many studies. Such studies have often been undertaken by groups who were early to recognize the power of this technique in biological applications but who had not yet achieved the necessary technical expertise to apply the technique successfully. Thus, in the heady days after the introduction of the technique by Liang and Pardee in 1992 (2), many biologists with divergent areas of expertise (neurobiology, psychology, plant sciences, pharmacology) rushed into applying the technique in their particular systems without, perhaps, a sufficiently solid background in modern molecular biology. This was a testimonial to the apparent simplicity of the technique and to the potential breadth of biological systems to which the technique might be applied.

People working in many different fields, and who had been interested in looking at changes in gene expression, suddenly had a technology that appeared to be accessible to virtually any reasonably well equipped biological laboratory. Many scientists had tried previous technologies to investigate differential gene expression such as subtractive hybridization and related techniques with mixed success. The apparent ease with which differential display could be performed caught the imagination of many. It soon became apparent, however, that while the potential breadth of application for differential display had been underestimated, the technical difficulties of the technique had clearly been underestimated as well. It struck participants at the first International Conference on Differential Display and Related Techniques (held in Halifax in 1995 and organized by the editors of this book and by Drs Liang and Pardee) that there was a tremendous enthusiasm for this type of technology to allow workers in

1

many areas of biology to investigate differential gene expression in their laboratories.

As we have seen, differential display initially had many problems. This book in the 'Practical Approach' series is a distillation of the successes that about a dozen laboratories have had in overcoming these difficulties and it provides a manual for the application of the technique to a wide variety of biological problems, with a major emphasis on the field of molecular neurobiology. A common complaint heard amongst differential display users is that they would like to be able to use non-radioactive labelling in the procedure. In Chapter 2, Karen Lowe describes efforts to develop differential display techniques based on the use of fluorescent probes.

In the following chapter, Andrew Medhurst and his colleagues deal with a common problem experienced by users of the differential display technique. Differential display was initially introduced as a means of comparing gene expression in a single immortalized cell line *in vitro* under different conditions (2). Attempts to apply differential display to heterogeneous tissues have, however, not always been successful. None the less some researchers have been successful in applying the technique in complex tissues. Medhurst and colleagues discuss in some depth the issue of experimental design for differential display of transcripts obtained from complex tissues.

An example of one of the considerations that must be addressed is illustrated by the following. A particular physiological or other challenge may lead to massive changes in gene expression in one cell type in a brain region or other organ but the remainder of the cells may remain unchanged. Such expression changes may be confined to a cell type that constitutes only a small fraction of the cells in that tissue such that the change in gene expression in those cells becomes lost in the background of unchanged gene expression in all the other cell types. In order not to miss important changes in gene expression in small cell populations, careful experimental design is necessary. Designing an experimental programme to maximize the chances of obtaining valid results is one of the most difficult parts of performing a successful differential display study. In essence, however, this is really a matter of applying common sense and good general experimental design. For example, if one is searching for changes in gene expression in the tiny suprachiasmatic nucleus of the brain to investigate the molecular biology of circadian rhythm, it is not a good plan to isolate RNA from the whole brain or a large subregion such as the ventral hypothalamus (which contains the suprachiasmatic nucleus). In fact it is necessary to start with RNA isolated from only the suprachiasmatic nucleus itself. Using these techniques to study changes in gene expression in such minute brain subnuclei is a real challenge, and new methods of carrying out differential display on small amounts of tissue are being developed to cope with this.

Sometimes the challenge for the molecular biologist is to determine exactly which tissues ought to be studied in the first place. This may be a simple matter; e.g. if one is interested in studying the gene changes in the brain that may be associated with Parkinson's disease, the substantia nigra is an obvious choice of tissue. If the aetiology of a disease is more poorly understood, however, the choice may be less obvious.

Several alternatives to 'classical' differential display have been developed. For example, RAP-PCR was developed at the same time as differential display and is a closely related technique using two random primers in the PCR amplification rather than the anchor primer approach. In Chapter 4, McClelland and Welsh and their colleagues describe recent developments in the use of this technique and, in particular, they describe expression profiling using reduced complexity probes for cDNA arrays. Nuccio and colleagues in Chapter 5 discuss the use of differential display in single-celled organisms and plant tissues.

Differential display has been improved in many ways. The description 'differential dismay' referred to difficulties that researchers had in the early days of the technique. Most people would agree that the polymerase chain reaction (PCR) is probably the most important molecular biological tool of the last quarter century. No one insists, however, that it be done using exactly the same technique as originally described by Mullis and his colleagues (3), as many elaborations and refinements in PCR have been developed. In the same way, numerous modifications have been made to the technique of differential display. These include major changes such as those described by Yerramilli and Weissman in Chapter 6. That technique, described as '3' end cDNA restriction fragment display analysis', was developed by Prashar and Weissman (4) and is based on the use of restriction enzymes to cut double-stranded cDNAs after the reverse transcription step. The refinement has some important theoretical advantages. The cDNA prepared is digested with a six nucleotide recognizing restriction enzyme and the products are ligated to the Y-shaped 'Fly-Adapter' and then subjected to PCR amplification. The PCR amplified products are subsequently analysed by electrophoresis on sequencing gels and the bands are visualized by autoradiography. In Chapter 7, Robertson and his colleagues describe another approach to differential display, which involves modification of a number of steps in the original technique. Yet another new approach to the study of differential gene expression is the technique of suppression subtractive hybridization, described by Wang and Feuerstein (Chapter 8). This is a PCR-based technique that depends on ligating adapters to each end of cDNAs. This technique combines many of the strengths of subtractive hybridization with the power of PCR.

Amongst these newer techniques for the analysis of gene expression, serial analysis of gene expression (SAGE) has attracted a great deal of attention (see the SAGE advice home page in ref. 5). Erno Vreugdenhil and his colleagues (Chapter 9) describe their approach to SAGE analysis and the application to the brain. There is much discussion of DNA and oligonucleotide array technologies, either using membranes or silicon 'chips' as substrates, as pointing the way to the future of differential gene expression studies. Arthur Pardee (6) has described the race between the older techniques, such as differential display, and the newer techniques, such as the DNA arrays, as a tortoise and hare race. Differential display is the tortoise. The real question, Pardee says, is 'Will cottage industry, using individual identification methods such as differential display, discover the majority of useful genes before parallel arrays get fully up to speed?' These 'hare' technologies are discussed in Chapter 10 by Debouck and her colleagues: 'The future of differential gene expression technologies: high density oligonucleotide microarrays and quantitative differential gene expression'.

Differential display and other gene screening techniques can be used together with techniques for knocking-down or otherwise regulating gene expression. One of the most exciting potential uses of differential display is in conjunction with transgenic or mutant strains of experimental animals. An excellent example of this is the discovery of the role of melanocyte stimulating hormone in the obese mouse (7). Wang and Uhl in Chapter 11 describe their use of differential display in the analysis of brain signalling pathways and the use of antisense oligonucleotide knock-down of gene expression to confirm functional significance.

In general, once we know that a given biological effect can be prevented by protein synthesis inhibition, there is good reason to suppose that the effect is regulated at the level of gene expression. The question then becomes one of 'expression of which gene?' This is an obvious instance of a biological question that could be addressed using differential display or similar techniques. It has long been known that protein synthesis inhibitors, for example, can block learning and memory. In Chapter 12, Huang and Lee describe their experiments on learning and memory in rats using differential display.

By the end of 1999 there will be more than a million expressed sequence tags (ESTs) identified, representing perhaps 50% of the estimated 140 000 genes in the human genome. Sequencing technologies are advancing so rapidly that the entire human genome will be sequenced within the next few years. The consequence of this pace of genomic data production is that the fraction of known genes that have had their biological functions elucidated is declining rapidly as our ability to sequence genes has outrun our ability to assess their physiological function. Techniques for investigating differential gene expression, such as differential display and the other techniques described here, will continue to be of crucial importance in this next stage of molecular biology.

References

1. Debouck, C. (1995). *Curr. Opin. Biotechnol.*, **6**, 597.
2. Liang, P. and Pardee, A. B. (1992). *Science*, **257**, 967.
3. Mullis, K. and Faloona, F. A. (1987). *Methods in enzymology*, **155**, 335.
4. Prashar, Y. and Weissman, S. M. (1996). *Proc. Natl. Acad. Sci. USA*, **93**, 659.
5. http://genome-www.stanford.edu/Saccharomyces/help/querySAGE.html#links
6. Pardee, A. B. (1997). *Nature Biotechnol.*, **15**, 1343.
7. Qu, D., Ludwig, D. S., Gammeltoft, S., Piper, M., Pelleymounter, M. A., Cullen, M. J., *et al.* (1996). *Nature*, **380**, 243.

Chapter 2
Recent advances in fluorescent differential display

Beckman Coulter, Inc., 1050 Page Mill Road, Palo Alto, CA 94308, USA

1 Introduction

Differential display (DD) was first described by Liang and Pardee (1) in 1992 as a method for analysing gene expression in eukaryotic cells and tissues. DD has been widely applied to study changes in mRNA expression induced by temporal development, diseases, and various cellular factors (2–7). This simple and powerful technique simultaneously screens for both up-regulated and down-regulated transcripts in multiple cell populations. To understand the advantages of DD, it is helpful to briefly review some other gene expression analysis methods.

1.1 Subtractive hybridization (SH)

This method (8) compares qualitative differences in gene expression between two cell populations by isolating the mRNAs which are uniquely present in one cell type. The typical SH process begins with the two cDNA libraries from the cell populations, and the synthesis of biotinylated first strand cDNAs from each species of mRNA present in the first ('driver') cell population (9, 10).

To obtain mRNAs that are up-regulated in the second ('target') cell population, the biotinylated driver cDNAs are prepared at many-fold molar excess over the cDNAs in the target cells, and are hybridized to the mRNA pool from the target cells. The duplexes of driver cDNAs and target mRNAs are then removed via binding of the biotin moieties to streptavidin, resulting in a pool of target mRNAs preferentially enriched in transcripts expressed only by the target cell population. Several rounds of mRNA:cDNA hybridization are required to remove the abundant mRNAs common to both cell populations. Any remaining unhybridized driver cDNAs are removed by binding these biotinylated species to streptavidin.

To determine the down-regulated mRNAs in the second cell type, the SH must be carried out in reverse, with biotinylated cDNAs synthesized from the second cell type now serving as the driver, and the cDNAs are hybridized to the mRNAs of the first cell population. Again, several rounds of hybridization with biotinylated driver cDNAs present in many-fold molar excess over the target mRNAs must be performed.

SH is effective for studying differential gene expression, but separate experiments are required to find up-regulated and down-regulated genes, respectively (11, 12). Furthermore, SH is practical for only pairs of samples (13, 14), such as a normal cell type and a single experimental cell type. SH requires significant mass amounts of mRNA for the multiple rounds of hybridization. Clearly, SH is not configured for screening multiple samples simultaneously.

1.2 Differential display (DD)

Instead of attempting to remove all but the differentially expressed mRNAs, DD converts poly(A)$^+$ mRNAs into first strand cDNAs via reverse transcription (RT) using oligo(dT)-anchored primers (APs), followed by polymerase chain reaction (PCR) using the 3′ 'downstream' APs in pairwise combination with 5′ 'upstream' arbitrary primers (ARPs). The cDNA fragment patterns from the control and experimental samples are then 'displayed' side by side on a high resolution denaturing gel so that differential gene expression is immediately apparent. The cDNA fragment pattern can be viewed as a 'bar code' for the expression profile of a particular eukaryotic cell population under the control and test conditions.

DD offers several advantages over other screening methods (15). Unlike SH, DD can examine several different RNA samples simultaneously, so DD is ideal for time course, dose–response, and multiple treatment studies. Furthermore, DD measures semi-quantitative differences in gene expression, rather than the complete absence or presence of different mRNA species. DD is a robust technique: the isolation of mRNA is not required, and comprehensive analyses consume only about 15 micrograms of total RNA from each cell population.

1.3 Random arbitrarily primed (RAP)-PCR

Welsh, McClelland, and colleagues (16, 17) developed a PCR-based approach using combinations of only ARPs. In RAP-PCR, because no 3′ oligo(dT) AP is employed, there is less representation of 3′ sequence. This may be an advantage when dealing with shorter cDNA fragments (< 500 nt), as they may represent portions of the protein coding sequence rather than the 3′ untranslated sequence. RAP-PCR can also screen transcripts that are not polyadenylated, such as prokaryotic mRNAs.

However, there are definite advantages of DD over RAP-PCR for eukaryotic model systems. If total RNA is the starting substrate in RAP-PCR, there is no way to confine the arbitrary priming to only the mRNA fraction. Since ribosomal and transfer RNAs are much more abundant than mRNAs in total RNA samples, most of the resulting cDNAs in RAP-PCR will derive from non-mRNA templates, and the RAP-PCR gel may contain a high percentage of non-relevant data. The requirement for purified poly(A)$^+$ RNAs in RAP-PCR may present technical difficulties in accurate quantification of the mRNA templates, or if residual oligo(dT) fragments contaminate the poly(A)$^+$ mRNAs and compete against the RAP-PCR primers during PCR, or if poly(A)$^+$ mRNA isolation alters the relative amounts of the different mRNAs.

1.4 Representational difference analysis (RDA)

RDA is a PCR-coupled genome subtractive process (18) that has been modified for use with cDNA (19). In cDNA RDA, the cDNA is digested with a tetranucleotide recognition sequence restriction enzyme, ligated to linkers, amplified by PCR, and the tester fragments are ligated to new, different linkers. Both the tester and driver fragments are digested, mixed together at ratios of tester: driver ranging from 1:100 to 1:800 000, hybridized, and selectively amplified. Only tester:tester hybrids will undergo exponential amplification. In practice, many rounds of hybridization and several different restriction enzymes are used in the RDA process.

The RDA subtractions are usually performed in forward and reverse directions in order to find both up- and down-regulated products, respectively. The enriched target bands are then analysed on an ethidium bromide stained agarose gel. The lower level of band resolution on the RDA agarose gel can lead to difficulties in obtaining pure fragments for each repetition of ligation, hybridization, and PCR.

DD is much faster and easier than RDA. Each of the hybridization periods in RDA may be as long as 20 hours, and the linker ligations may each require 12–14 hours. Like SH, RDA requires hundreds of micrograms of each RNA sample, so RDA is untenable if the RNA source is limiting. Similar to SH, RDA is better suited to pairs of matched control and test samples, rather than to time course, dose–response, or multiple treatment comparisons.

1.5 Subtractive display (SD)

The SD process (20) is carried out by isolating mRNA from control and test cells or tissues for cDNA synthesis, subjecting the cDNAs to restriction enzyme digestion, linker ligation and PCR amplification, and then removing cDNAs that are expressed in common via repeated rounds of subtraction during alternating long (20 hours) and short (2 hours) hybridization periods. The subtracted cDNA products are selectively PCR-amplified and resolved or displayed on a sequencing gel.

The main advantage of SD is the simplified display pattern, since the bands common to both cDNA populations will be eliminated. The disadvantages of SD are similar to those of SH and RDA: the procedure is cumbersome if both up- and down-regulated genes are of interest, particularly if several samples are to be compared. SD is not well suited to dose–response, time course, or multiple test condition studies. Like SH and RDA, SD requires considerable mass amounts of RNA to carry out the repetitive hybridizations. With hybridization-based methods, the stringency of the hybridization conditions must be optimized empirically for each set of samples. Also, there are many steps involved in the restriction enzyme digestions, linker ligations, and numerous long- and short-term hybridizations, all preceding the PCR amplification and display.

DD is much more streamlined, and the DD gel pattern is rapidly analysed by simply focusing on cDNA bands with unique or significantly greater signals either in the control samples or in the experimental samples. High resolution gel matrix such as HR-1000 (Beckman Coulter, Inc.) and temperature controlled electrophoresis systems such as the GenomyxLR DNA Sequencer (Beckman Coulter, Inc.) have successfully improved the DD band patterns by providing greater separation and reproducible, uniform mobility of the cDNA fragments (21, 22).

Furthermore, since the cDNAs are anchored at the 3′ end and are not restriction digested in DD, the larger DD-PCR products are more likely than shorter DD-PCR products to contain coding information. In contrast, it is more difficult to predict the potential utility of an SD band from its size.

2 Differential display strategy and primer design

DD uses two types of primers to convert subsets of the mRNA pool into double-stranded cDNAs, which are electrophoretically resolved to generate an expression profile 'bar code'. The first type of primer, present in first strand cDNA synthesis RT reactions and serving as the 3′ or 'downstream' primer in DD-PCR, is the 'anchored' primer (AP). Most APs contain a string of 10–12 dTs that can anneal with the poly(A)$^+$ tail of the mRNAs. APs employ either one or two 'anchoring' bases at the 3′ end to anneal to specific subsets of mRNAs. First strand cDNA synthesis catalysed by RT occurs off the AP, thereby *anchoring* the 5′ end of the cDNA transcript near the start of the poly(A)$^+$ mRNA tail. Single base (A, C, and G) anchored oligo(dT) primers generate three different fractions, corresponding to the presence of either a U, G, or C respectively, immediately upstream of the poly(A)$^+$ region. Assuming that a given cell type expresses about 10 000–15 000 different species of mRNA, each single base anchored AP would be expected to generate about 3000–5000 different first strand cDNAs. Such a large number of different cDNAs would be difficult to resolve into distinct, homogeneous bands on any sequencing gel.

There are 12 possible different two base combinations (reading 5′ to 3′) immediately upstream of the poly(A)$^+$ tail (UC, GC, CC, AC, UG, GG, CG, UU, GU, CU, AU, and AG). Each corresponding oligo(dT) AP (respectively oligo(dT) GA, -GC, -GG, -GT, -CA, -CC, -CG, -AA, -AC, -AG, -AT, and -CT, reading 5′ to 3′) would hybridize to about one-twelfth (8%) of the mRNA pool, and would generate about 800–1200 different first strand cDNAs. Clearly, each of the 12 different oligo(dT) APs, employing two anchoring bases, will produce a more manageable, resolvable number of cDNA fragments than any of the three single base oligo(dT) APs. *Figure 1* depicts the annealing of a two base anchored oligo(dT) AP with an appropriate poly(A)$^+$ mRNA.

The first strand cDNA synthesis is catalysed by RT enzyme. The optimized Genomyx HIEROGLYPH™ mRNA Profile Kit System (Beckman Coulter, Inc.)

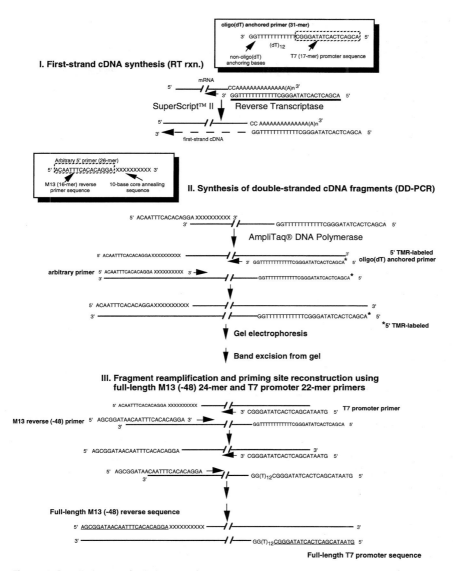

Figure 1 Detailed view of priming reactions.

specifies the use of SuperScript™ II RNase H⁻ RT (Gibco BRL Life Technologies) to enhance the yield of longer cDNAs. Longer, intact cDNAs are preferred because they make the DD screening and downstream analyses more efficient. As the length of a first strand cDNA approaches full-length, the probability increases that a greater number of arbitrary priming sites will be present on the template. The probability that a particular AP and ARP pair used in DD-PCR can produce a double-stranded cDNA fragment from a particular mRNA, for display on the DD gel, also increases.

The cDNA fragments synthesized in DD are typically rich in sequence derived from the 3' end of the mRNA, because the APs anneal to the upstream end of the poly(A)$^+$ tail. Smaller fragments ($<$ 300 nt) may be almost exclusively composed of untranslated sequence which is not very useful for cDNA database comparisons with sequences derived primarily from the coding portion of the gene. It is therefore preferable to produce and select longer cDNAs, which are more likely to include informative protein encoding sequences.

The HIEROGLYPH™ mRNA Profile Kit System has been optimized to produce cDNAs of up to 1.5–2 kb with virtually perfect reproducibility between duplicate samples. These larger cDNAs provide more coding information which is useful for the identification and functional characterization of the mRNA. Most important, high reproducibility is absolutely required for reliable DD. The HIERO-GLYPH mRNA Profile System is composed of five separate mRNA Profile Kits for the comprehensive, systematic analysis of eukaryotic mRNA expression. Each Profile Kit contains all 12 anchored 3' primers (AP1–12), in combination with 4 out of 20 different arbitrary 5' primers (ARP1–20).

2.1 HIEROGLYPH™ oligo(dT) anchored primers

HIEROGLYPH downstream APs use two anchoring nucleotides at the 3' end of the primer immediately following the 12 nt oligo(dT) region. This design ensures more efficient and reproducible screening of target mRNA subpopulations by generating a more manageable number of first strand species. Each AP is 31 nt in length and incorporates a 17 nt T7 RNA polymerase promoter-derived site which permits directional sequencing from the 3' end of the cDNA without the necessity for subcloning (*Figure 1*). The T7 promoter also allows an antisense RNA transcript to be generated for verification of differential expression of the cDNA fragment via a ribonuclease protection assay (RPA), *in situ* hybridization, or Northern analysis.

After first strand cDNA synthesis is completed, aliquots of these RT reactions are exponentially amplified by PCR using the original AP in combination with an upstream ARP, converting the first strand cDNAs into double-stranded cDNAs.

2.2 HIEROGLYPH arbitrary primers

The HIEROGLYPH System employs 20 different ARPs, each of which contains a 10-mer core annealing sequence, as seen in *Figure 1*. Each ARP is 26 nt in length and incorporates a 16 nt segment of the M13 reverse (–48) 24-mer priming sequence, which allows directional sequencing of the coding strand of the cDNA.

The 10-mer core arbitrary sequences behave as 6-mers, 7-mers, and 8-mers, leading to more frequent annealing, and the synthesis of more cDNA fragments than would occur if all ten bases were involved in priming (the probabilities of a 6-, 7-, or 10-mer finding a complementary annealing sequence are one in 4^6, one in 4^7, and one in 4^{10}, respectively). This flexibility occurs because different numbers of bases within the 10-mer sequence can establish annealing sites during the initial, moderately stringent cycles of DD-PCR. After the first four cycles, the annealing temperature is increased to raise the stringency and specificity of the DD-PCR (7).

This arbitrary priming flexibility is beneficial for DD analysis. First, it increases the chances that a particular ARP will anneal to the template and synthesize a cDNA fragment, thereby increasing the mRNA representation coverage. Secondly, it allows the ARPs to overlap slightly to reduce the possible gaps in coverage. The PCR conditions are selected for priming plasticity of the 10-mer core sequences, but with sufficient control to avoid synthesis of non-specific cDNAs. A set of twenty 10-mer core arbitrary sequences, used under the appropriate PCR conditions, should be sufficient to ensure coverage of virtually all mRNAs in a eukaryotic RNA sample.

The native sequences of both the T7 promoter and M13 reverse (–48) in the APs and ARPs, respectively, were shortened to avoid excessively long and unwieldy primers. *Figure 1* depicts how the full-length M13 reverse (–48) and T7 promoter sequences are regenerated during target band reamplification.

2.3 The importance of cDNA fragment length

A vital feature of the HIEROGLYPH DD process is its ability to synthesize cDNA fragments which are much longer than those generated by some other DD approaches. When one evaluates a DD gel, one sees a spectrum of different sized cDNAs. With the HIEROGLYPH System, cDNAs at the upper end of this size spectrum are approximately 1.5–2 kb in length, or roughly twice as large as those generated by conventional systems. The ability to reliably synthesize longer fragments has several important consequences.

(a) By shifting the size spectrum upward, there is an increase in the amount of coding information which each cDNA fragment may include. For differentially expressed cDNAs of > 800–1000 nt, the coding information capacity is significantly greater than that of cDNAs in the conventional 100–500 nt size range.

(b) The ability to synthesize and resolve longer cDNAs increases the comprehensiveness of transcript coverage by allowing the first strand cDNAs that are arbitrarily primed far upstream of the anchored priming site to be represented on the DD gel. The HIEROGLYPH cDNA fragments displayed on the DD gel tend to be longer (more informative) and less numerous (less complex and less redundant). Such DD patterns, resolved on the high resolution HR-1000 gel matrix and GenomyxLR electrophoresis system, are more easily and rapidly interpreted.

(c) These longer fragments are much more useful as probes in confirmatory steps such as RPA, *in situ* hybridization, or Northern analysis, because of the higher specific activity and the higher stringency hybridization conditions which can be achieved with longer antisense RNA probes. The confirmation of differential gene expression becomes more reliable, sensitive, and specific. These advantages are key to successful DD projects.

3 Fluorescent differential display (fluoroDD)

A further advancement in DD has been the introduction of faster, safer, and more cost-effective methods that avoid the use of radioactivity: fluorescent analysis of the cDNA band patterns via the optimized fluoroDD Kit in conjunction with the HIEROGLYPH mRNA Profile System, the GenomyxLR™ DNA Sequencer electrophoresis system, and the GenomyxSC™ Fluorescent Imaging Scanner (Beckman Coulter, Inc.). The fluoroDD kit provides all 12 APs pre-labelled with a tetramethylrhodamine (TMR) fluorescent 5′ tag, and is designed to be used in the DD-PCR stage in combination with any of the five HIEROGLYPH mRNA Profile kits.

The fluoroDD procedure using Kit 1 of the HIEROGLYPH mRNA Profile System is presented schematically in *Figure 2*. The RNA samples are templates for

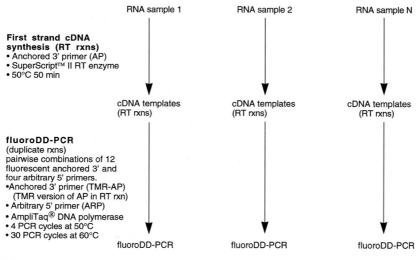

Figure 2 Overview of the fluoroDD mRNA profile procedure.

first strand cDNA synthesis in RT reactions using the 12 different HIEROGLYPH oligo(dT) 3′ APs. The cDNAs are then used in duplicate fluoroDD-PCR reactions with the TMR-labelled APs used in pairwise combinations with the different 5′ ARPs. Each RNA sample is amplified by 48 primer–pair combinations (96 duplicate reactions per RNA sample) from Kit 1, as shown in *Figure 3*. The duplicate fluoroDD-PCR samples are loaded in a consistent order on adjacent lanes of a

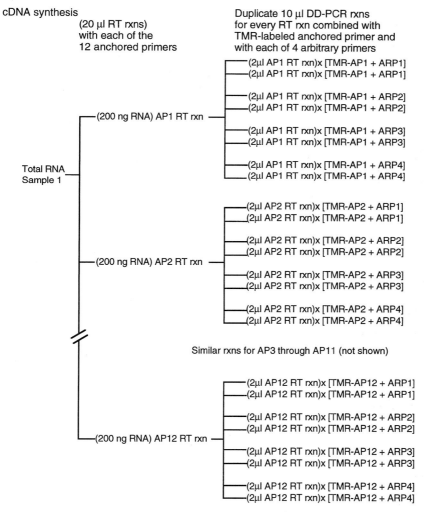

Figure 3 Flowchart of RT and fluoroDD-PCR strategy. fluoroDD is shown with HIEROGLYPH Kit 1 and RNA sample 1. In practice, each of the RNA samples (1 through N) would be set up in parallel reactions with the same TMR-AP and ARP primer pair combinations, so that the control and experimental fluoroDD-PCR samples corresponding to the same primer pair can be analysed in adjacent gel lanes. The researcher may opt to work systematically with a subset of the anchored primers at a time in combination with all four ARPs of the kit, depending on the number of RNA samples to be compared, in order to streamline sample processing and analysis.

high resolution denaturing gel, with reactions from different RNAs, amplified with the same AP and ARP pairs, grouped together in consecutive lanes.

Using all five HIEROGLYPH kits with the fluoroDD kit for comprehensive transcript coverage involves 12 AP × 20 ARP × 2 duplicates or 240 PCR reactions for each control RNA and each experimental RNA sample. In practice, the researcher works systematically with a subset or kit of primers at a time, setting up all of the RNAs to be compared in parallel reactions with the same AP and ARP combinations, so that all of the control and experimental fluoroDD-PCR samples generated with the same primer pair can be analysed in adjacent gel lanes.

The optimized fluoroDD-PCR protocol minimizes the formation of non-specific products and 'false positives', and ensures high reproducibility between samples. The fluoroDD kit, in conjunction with the HIEROGLYPH mRNA Profile System, provides the researcher with a high degree of confidence in the DD-PCR patterns. Furthermore, the TMR tag on the fluoroDD APs allows rapid and sensitive detection of the DD patterns by the GenomyxSC fluorescent imaging scanner. The fluorescent gel can be scanned in about an hour, in contrast to an overnight autoradiogram in classic radioactive DD.

The fluorescent cDNA band patterns are easier to interpret than radioactive DD patterns due to lower background and fewer doublet bands (*Figure 4*). Doublet bands are seen on autoradiograms because both coding and non-coding cDNA strands are radiolabelled during PCR incorporation of $[\alpha\text{-}^{33}P]dATP$. During high resolution gel electrophoresis, the radiolabelled cDNA strands are denatured, and they migrate as distinct bands or doublets. The more complicated radiolabelled banding patterns may result in additional efforts expended needlessly in pursuit of the doublet bands, which are complementary strands of the same cDNAs.

Since the fluorescent tag is present only on the APs, the observed fluoroDD-PCR bands most likely represent products generated from the poly(A)$^+$ mRNA.

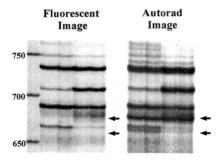

Figure 4 Comparison of fluorescent DD gel pattern with radioactive DD autoradiogram. Due to the difference between TMR-AP primer end-labelling (fluoroDD) and direct $[\alpha\text{-}^{33}P]dATP$ incorporation (radioactive DD), the fluoroDD gel band patterns have lower background and fewer doublet bands compared to the radioactive DD pattern of the same RNAs generated from the same primer pairs.

The signal intensity of the TMR-labelled band is also more representative of the actual number of cDNA molecules present in the band, since there is only one tag per primer molecule. Radioactive incorporation tends to give a stronger signal for longer cDNA products compared to shorter cDNA products, because a longer cDNA product will incorporate a greater number of radiolabelled nucleotides, and thus radioactive DD tends to overestimate the amount of cDNA in a high molecular weight band. Another convenient feature of fluoroDD samples is their stability: they can be prepared in large batches and stored at 4°C or −20°C, securely wrapped in aluminium foil to protect them from light, for at least four weeks before they are analysed on 5.6% HR-1000 clear denaturing gels. This makes fluorescent DD more amenable to automation and higher throughput sample processing.

3.1 Coverage with the fluoroDD system

The ability to generate and visualize the fluoroDD-PCR products depends on several factors, such as the purity and integrity of the total RNAs, the relative abundance of a particular transcript, the priming efficiency, the resolving power of the electrophoresis system, and the sensitivity of the imaging system. As described earlier, the HIEROGLYPH AP design used by the fluoroDD system reduces the number of first strand cDNAs synthesized in any given RT reaction. This simplification of the cDNA fragment subpopulations facilitates more thorough and accurate screening by offering the first strand cDNAs adequate access to primers and other reagents during fluoroDD-PCR, resulting in their semi-quantitative representation on the fluorescent DD gel (23).

Conventional 45 cm long sequencing gels generally cannot resolve DD-PCR products that are longer than 500 nt, so complete detection of all DD-PCR species is virtually impossible with conventional sequencing systems. The GenomyxLR DNA Sequencer routinely resolves discrete DNA bands up to about 1.5–2 kb, and is highly recommended for fluoroDD as well as classic HIEROGLYPH DD analysis.

3.2 A note on cDNA fragment expression patterns

Since the 12 APs are designed to specify different subpopulations of the mRNA pool, the number, length, and distribution of cDNA fragments will vary when different AP and ARP combinations are used. In addition, all upstream nucleotide combinations are probably not equally represented in a particular RNA population (24). For example, a particular ARP may produce a great number of cDNA fragments when used with the APs with a 'G' in the penultimate position, but may generate a relatively low number of bands when used with APs having a penultimate 'A'. Also, certain AP-ARP combinations may give relatively large cDNA fragments, while other combinations may result in somewhat shorter species. It would be difficult to predict which of the AP-ARP combinations will generate the majority of differentially expressed cDNA fragment(s) or the longest cDNA products for a given RNA sample.

3.3 **Proper experimental design and RNA sample selection**

It is essential to design a suitable, properly controlled DD project to minimize any confounding factors. For example, a well-designed study might compare the induced and uninduced transcripts of one particular cultured cell line or of one carefully controlled model system in response to two or more cellular conditions or treatments. If possible, it is better to analyse RNA samples from homogeneous cell populations, because it is more difficult to replicate the results obtained from heterogeneous cell populations. Comparisons of profiles generated from whole tissues may give rise to false positives and irreproducible results. The entire set of total RNA samples being compared in a DD experiment should be purified in parallel, using the same RNase-free reagents and methods. The control and experimental RNAs must be treated with RNase-free DNase I to remove genomic DNA that can lead to artefacts and false positives.

It is important to make sure that all pipetters used in RNA sample purification and DD experiments are carefully calibrated for accuracy and precision. Always use aerosol-resistant, sterile, nuclease-free pipette tips. Be sure to use high quality, uniform, thin-walled PCR reaction tubes, as specified in the protocols. It is a good practice to dedicate completely separate sets of pipetters for pre-PCR and for post-PCR work, in order to prevent contamination of starting materials and reagents with amplified products. If possible, perform all pre-PCR steps in a clean area of the laboratory, such as a laminar flow hood, separate from post-PCR analysis areas.

Confirmation of differential gene expression should be carried out with the appropriate RNA samples generated from an *independent* experiment. This is a logical but sometimes overlooked detail in the design of differential display projects. This independent experiment should be identical to the original experiment, but carried out separately from the original DD study. Successful confirmation of differential gene expression is most often achieved (at a lower false positive rate) when a sensitive assay method is used, such as Northern blot analysis, RPA, or *in situ* hybridization. In cases where the amount of sample RNA is extremely limited, the reverse Northern method may be used (22). Also, the experimental model system may contribute significantly to the ease of confirming differential expression: in general, cell lines are often more amenable because the cell population is more uniform and controlled. Sequencing of the candidate bands is recommended as a step prior to the confirmation because it is a rapid method and can help in determining whether the gene is of interest for further characterization (22).

3.4 **Preparation of total RNA samples**

The purity and integrity of the total RNA samples are the most vital factors for successful DD analysis. Proper laboratory practices must be followed to ensure the quality of the RNAs and to avoid contamination and degradation by RNases. RNases may be introduced inadvertently through improper technique; since these enzymes are difficult to inhibit or to denature irreversibly, it is extremely important to guard against introducing them whenever RNA samples are prepared or handled. The following guidelines are strongly recommended:

(a) Always wear clean, powder-free disposable gloves (UL-315-S/M/L, Microflex), since skin, body fluids, bacteria, etc., can be sources of RNases. Change the gloves as often as necessary.

(b) Practice good microbiological technique to prevent contamination of the materials, reagents, and equipment used to prepare and analyse RNA samples.

(c) All reagents and equipment must be kept as nuclease-free as possible. Store all materials and reagents to be used for RNA work in a restricted-use area, and mark them with labels: 'For RNA work only! Keep nuclease-free! Handle only with nuclease-free implements and gloved hands!'

(d) Use sterile, nuclease-free, disposable plasticware (Ambion) and automatic micropipetters reserved for RNA work.

(e) Rinse the removable micropipette barrels with bleach or 0.5 M NaOH, and then rinse thoroughly with sterile, nuclease-free water. Autoclave the cleaned and dried pipette barrels before use.

(f) Always use sterile, nuclease-free pipette tips with aerosol-resistant filters, to minimize the risk of cross-contamination. Keep the pipette tip boxes covered when not in use.

(g) Any glassware or metal utensils to be used in RNA sample preparation or handling should be washed carefully, rinsed thoroughly with sterile, nuclease-free water, wrapped with aluminium foil, and baked at 150°C for at least 4 h.

(h) Use protective equipment and safety precautions whenever handling hazardous reagents.

Total RNA samples should be prepared from cultured cells or homogenized tissues using a standard guanidinium isothiocyanate (GITC)/ acidified phenol/ chloroform method (25) or a reliable commercial kit according to manufacturer's instructions. In cases where the cell or tissue type may have high levels of endogenous RNases, it is advisable add 2-mercaptoethanol to the GITC lysing buffer to promote inactivation of RNases. Note that the use of RNase inhibitor (e.g. N2511, Promega; 2682, Ambion) is optional as long as extreme care has been taken to avoid introduction of RNases during RNA sample preparation. Indeed, RNase inhibitors cannot substitute for good laboratory practices, since the commercially available RNase inhibitors do not inactivate RNase T1 or bacterial or fungal RNases. Furthermore, RNases can be released from the inhibition complexes in the event of oxidation or denaturation. Do not use vanadyl RNase inhibitors, which can interfere with subsequent reactions.

The use of purified poly(A)$^+$ mRNA in place of total RNA is unnecessary for DD, and may be undesirable because of possible contamination and interference by oligo(dT), leading to high background signals. Furthermore, during poly(A)$^+$ isolation, there may be potential loss of rare mRNA species.

Protocol 1

DNase treatment of total RNA samples

Equipment and reagents[a]

- Sterile, nuclease-free 1.5 ml microcentrifuge tubes (12400, Ambion)
- Water-bath at 37 °C
- Floating microcentrifuge tube rack
- Refrigerated microcentrifuge at 4 °C
- Dry ice/ethanol bath, −70 °C
- −70 °C freezer
- UV spectrophotometer and ultra-microcuvettes
- Horizontal slab gel electrophoresis unit
- UV transilluminator or GenomyxSC fluorescence scanning imager (Beckman Coulter)
- Total RNA aqueous samples, prepared using a reliable method[b] or kit (e.g. ToTALLY RNA™, 1910, Ambion; RNAzol™, CS-104 or RNA-STAT-60™, CS-110, Tel-Test; or TRIzol™, 15596-026, Gibco BRL)
- 1 M Tris–HCl pH 7.5, sterile, nuclease-free

- Sterile nuclease-free H_2O (E476, Amresco; 9920, Ambion; or DEPC-treated H_2O[c])
- 25 mM $MgCl_2$
- RNase-free DNase[d] I (27-0514-01, Pharmacia Biotech; or 2222, Ambion)
- Phenol:chloroform 1:1 (v/v) pH 5.2 (0966, Amresco)
- 8.0 M LiCl[e] (L7026, Sigma)
- Absolute ethanol, 70% ethanol prepared with sterile, nuclease-free H_2O
- 0.8% agarose gel in 1 × TAE buffer: 0.04 M Tris–acetate, 0.001 M EDTA
- Gel loading dye from the HIEROGLYPH mRNA Profile Kit (Beckman Coulter, Inc.)
- RNA standard molecular weight marker ladder (Ambion or Gibco BRL)
- SYBR Green II (S-7564, Molecular Probes) or ethidium bromide[f]

Method

1. Set up a DNase I digestion for each total RNA sample in sterile, nuclease-free, 1.5 ml microcentrifuge tubes by combining the following reagents in order:

Reagent	[Stock]	1 × vol. (50 μl)	[Final]
Total RNA, 25 μg	1 mg/ml	25 μl	0.5 mg/ml
Tris–HCl pH 7.5	1 M	2.5 μl	50 mM
$MgCl_2$	25 mM	20 μl	10 mM
RNase-free DNase I	2 U/μl	2 μl	0.08 U/μl

2. Mix the reactions gently by rotating the tubes end-over-end ten times (do not vortex). Spin the tubes briefly to collect the contents in the bottom of each tube. Incubate the reactions for 20 min at 37 °C in a water-bath.

3. Add 0.2 ml of phenol:chloroform pH 5.2, to each sample, vortex the tubes for 10 sec, and centrifuge them at 4 °C at maximum speed for 15 min. Transfer the upper, aqueous phase to a new sterile, nuclease-free tube. Avoid the interphase, and take care not to transfer any organic phase to the new tube.

4. To the upper phase, add 23 μl of 8.0 M LiCl (2.5 M final [LiCl]) and 0.2 ml of absolute ethanol. Mix by inverting the tube several times, and cool to −70 °C in a dry ice/ethanol bath for 10 min. Centrifuge the samples at 12 000 g for 15 min at 4 °C.

Protocol 1 continued

5. Carefully pipette off and discard the supernatant; carefully wash the RNA pellet twice with 0.5 ml of ice-cold 70% ethanol.[g]

6. Add 1 ml of 70% ethanol to each washed RNA pellet. Store the RNA pellets at −70°C.

7. When ready to assay the purity and integrity of the RNA samples, centrifuge the tubes at 12 000 g for 10 min at 4°C to obtain the RNA pellet at the bottom of tube. Carefully pipette off the 70% ethanol and allow the pellet to air dry for 10 min.[h] Redissolve the RNA thoroughly in 20 μl of sterile, nuclease-free water.

8. Dilute a 2.0 μl aliquot of the redissolved RNA to a final volume of 500 μl with sterile, nuclease-free water. Determine the OD_{260} and OD_{280} of the RNA.[i] (Heat the diluted RNA at 65°C for 5 min prior to taking the UV absorbance measurements, to ensure that the RNA secondary structure does not bias the readings.)

9. Electrophorese 1 μg of the redissolved RNA with gel loading dye on a 0.8% agarose, 1 × TAE gel at 80 V for 20 min. Stain the gel with SYBR Green II or ethidium bromide. Check the RNA bands on a UV transilluminator or a fluorescence scanner.[j]

10. If high quality RNA samples have been obtained, as assessed by steps 8 and 9, store the RNAs at −70°C in 5 μg aliquots (at a concentration of at least 1 mg/ml). For long-term storage, store the RNAs as ethanol precipitates at −70°C.

[a] All equipment and reagents must be kept RNase-free.

[b] See ref. 25.

[c] See ref. 26.

[d] Caution: DNase I is exceedingly sensitive to physical denaturation by shaking, vortexing, and vigorous pipetting. Mixing should be achieved by gentle inversion.

[e] LiCl precipitation of high molecular weight RNA is carried out on all RNA samples to remove low molecular weight cellular metabolites that can interfere with RT and DD.

[f] SYBR Green is approx. 25 times more sensitive than ethidium bromide for nucleic acid detection. Use care in handling these mutagenic compounds.

[g] Caution: the RNA pellet may be difficult to see.

[h] Important: be sure that no residual ethanol remains before redissolving the RNA pellet.

[i] The ratio of OD_{260} to OD_{280} should be greater than or equal to 1.9, and with the 250-fold dilution factor used here, the concentration of RNA in mg/ml is ten times the OD_{260} reading. (1 OD unit at 260 nm = 40 μg RNA/ml.)

[j] The total RNA should appear as a tight smear of approx. 500–2000 bp on the agarose gel. The 28S rRNA band at 4.7 kb should be about twice the intensity of the 18S rRNA band at 1.9 kb. If any trace of RNA degradation is observed, do not use the RNA for DD. Poor RNA quality can lead to poor differential display results and severe problems in downstream analyses.

After total RNA isolation, all of the RNA samples are treated with RNase-free DNase I, followed by phenol:chloroform (pH 5.2) extraction, and LiCl ethanol precipitation, to remove any genomic DNA. This DNase I treatment is crucial to the success of the DD screening. Even trace amounts of contaminating DNA can be amplified and may contribute false positives to the DD banding pattern. It is absolutely essential that the total RNA samples are completely free of DNA contamination. The purity and integrity of each DNase I-treated total RNA sample must be verified before proceeding with the first strand cDNA synthesis. The ratio of the optical densities at 260 nm and 280 nm should be greater than or equal to 1.9. The RNA samples should appear to be undegraded, as judged by their appearance on an agarose gel. If the samples do not meet the standard, new RNA samples should be purified and tested to ensure successful DD results.

3.5 First strand cDNA synthesis

Reverse transcription reactions (RT rxns) are performed with the HIEROGLYPH mRNA Profile Kit(s) to obtain 12 first strand cDNA subgroups from each RNA sample, using the 12 non-fluorescent oligo(dT) APs individually with aliquots of each total RNA sample.

A 20 μl RT rxn, using one of the 12 HIEROGLYPH oligo(dT) 3′ APs and 200 ng of one of the total RNA samples, will generate enough first strand cDNA for duplicate DD-PCR with the fluorescent version of the same oligo(dT) anchored 3′ primer in pairwise combination with all four HIEROGLYPH 5′ ARPs (eight fluoroDD-PCRs total) in a given HIEROGLYPH mRNA Profile Kit. Therefore, for any of the five HIEROGLYPH kits, 12 different 20 μl RT rxns will be carried out for each RNA sample. Each RT rxn will contain one total RNA sample and one of the 12 APs. First strand cDNA synthesis with all 12 oligo(dT) APs of a single HIEROGLYPH kit will consume 2.4 μg of each total RNA.

The first strand cDNAs should be stored at −70°C or −20°C in a constant temperature freezer, and used in DD-PCR within one week. It is usually more convenient to work systematically with a subset of the APs at one time, in combination with all four ARPs of the kit, rather than with all 12 APs all at once. Do not use cDNAs that have undergone more than one freeze–thaw.

Protocol 2

First strand cDNA synthesis

Equipment and reagents[a]

- 0.2 ml thin-walled MicroAmp (Perkin Elmer) PCR[b] tubes (N801-0580) and caps (N801-0535)
- The following thermal cyclers have worked successfully: Perkin Elmer GeneAmp 9600; Genomyx CycleLR thermal cycler (GX102) with a heated lid; or MJ Research PTC100 thermal cycler with a heated lid. Other thermal cyclers may require protocol modification.
- SuperScript™ II RNase H⁻ reverse transcriptase[c]: 200 U/μl (18064-014, 10 000 U; or 18064-071, 4 × 10 000 U) supplied

with 100 mM DTT and 5 × first strand cDNA synthesis buffer (250 mM Tris–HCl pH 8.3, 375 mM KCl, 15 mM MgCl$_2$) (Gibco BRL Life Technologies)
- Two or more total RNA samples,[d] high quality, free from DNA contamination (see *Protocol 1*), and prepared according to a properly designed experiment
- Genomyx HIEROGLYPH mRNA Profile Kit(s) (146081-85, Beckman Coulter, Inc.)
- Sterile nuclease-free H$_2$O (E476, Amresco; 9920, Ambion; or DEPC-treated H$_2$O[e])

Method

1. Thaw out the reagents (total RNA samples, unlabelled HIEROGLYPH APs, dNTP mix, HeLa RNA, 5 × SuperScript II RT first strand buffer, DTT) on ice, and keep them on ice. Mix each thawed reagent gently but do not vortex; spin the tubes briefly to collect the contents in the bottom of each tube.

2. Dilute the total RNA samples with sterile, nuclease-free water to a working concentration of 0.1 mg/ml. Dilute only as much RNA as needed for the RT rxns; diluted RNAs should not be refrozen for later use. Keep the tubes on ice.

3. Using the 0.2 ml thin-wall PCR tubes, prepare RT tubes for each total RNA sample (one RT rxn for each AP). Prepare a set of positive control reactions using the HeLa kit control RNA, freshly diluted to 0.1 mg/ml in nuclease-free H_2O, with each of the APs that are to be used.[f] Set up a negative control tube (–RNA control) containing nuclease-free H_2O in place of total RNA. For each experimental RNA sample (control and test conditions) prepare another negative control tube (–RT control) containing experimental RNA and no RT enzyme.[g]

4. In the appropriate tubes combine the following:
   - Total RNA, freshly diluted to 0.1 mg/ml   2.0 µl
   - HIEROGLYPH T7(dT$_{12}$)AP (2 mM)   2.0 µl

5. Mix the samples carefully with a micropipette (do not vortex), and spin the tubes briefly. Cap the tubes securely.

6. Incubate the RNA and AP at 70 °C for 5 min in a thermal cycler with a heated lid. Quickly chill the tubes on ice. Spin the tubes briefly.

7. On ice, set up the RT core mix in sufficient volume for the number of RT rxns. Also prepare a separate –RT core mix, substituting sterile nuclease-free H_2O for the RT enzyme, for use in the –RT control. Add 16 µl of core mix into the appropriate tubes for a final volume of 20 µl per tube. Mix the samples well with gentle pipetting. Cap all of the tubes securely. Spin the tubes briefly.

   (a) RT core mix.

Reaction component	[Stock]	1 × vol.	[Final]
Sterile nuclease-free H_2O	–	7.8 µl	–
SuperScript II RT buffer	5 ×	4.0 µl	1 ×
dNTP mix (1:1:1:1)	5 mM each	2.0 µl	0.5 mM each
DTT	100 mM	2.0 µl	10 mM
SuperScript II RT enzyme	200 U/µl	0.2 µl	2 U/µl

   (b) –RT core mix for –RT negative control tubes.

Reaction component	[Stock]	1 × vol.	[Final]
Sterile nuclease-free H_2O	–	8.0 µl	–
SuperScript II RT buffer	5 ×	4.0 µl	1 ×
dNTP mix (1:1:1:1)	5 mM each	2.0 µl	0.5 mM each
DTT	100 mM	2.0 µl	10 mM
SuperScript II RT enzyme	200 U/µl	–	–

Protocol 2 continued

8. Perform the RT rxns in a thermal cycler with a heated lid.
   (a) 42°C for 5 min.
   (b) 50°C for 50 min.
   (c) 70°C for 15 min.
   (d) Hold at 4°C.

9. Store in a −20°C constant temperature, nonfrost-free freezer for use in fluoroDD-PCR within one week.

[a] All equipment and reagents must be kept RNase-free. Note that the use of RNase inhibitor is less important than taking extreme care to avoid introduction of RNases during RNA sample processing and RT reaction set up. Be aware that the commercially available RNase inhibitors do not inactivate RNase T1 or bacterial or fungal RNases, and that RNases may be released from inhibitor complexes during oxidation or denaturation. Do not use vanadyl RNase inhibitors, which can interfere with subsequent reactions. If RNase inhibitor (2682, Ambion or N2511, Promega) is to be used, it must be added to all tubes: for each reaction add 1 μl of 20 U/μl stock to the RT core mix for a final concentration of 1 U/μl. Adjust the volume of sterile nuclease-free water accordingly.

[b] The polymerase chain reaction (PCR) is covered by US patents owned by Hoffman-LaRoche.

[c] Be sure to store this temperature-sensitive enzyme in a constant temperature (not frost-free) −20°C freezer. Leave the SuperScript II RT in the freezer until just before use and return the enzyme to the freezer immediately after use.

[d] At least 2.4 μg of each RNA sample will be required for the RT step (200 ng RNA with each of the 12 APs) to provide sufficient cDNA in duplicate DD-PCR reactions with the 12 oligo(dT) 3′ APs and the four 5′ ARPs, for a complete screening with the 48 primer-pair combinations of one mRNA Profile Kit.

[e] See ref. 26.

[f] The HeLa kit control RNA is a positive control to check the performance of the reaction. It is not an appropriate reagent for the –RT negative controls.

[g] Generally one AP can be used for the negative controls. The purpose of the –RNA control is to test the purity of the RT reaction components (there should be no products). The purpose of the –RT control is to test for genomic DNA contamination in the experimental RNAs that can contribute artefactual DD-PCR bands to the DD pattern.

The RT protocol given here is recommended for fluoroDD only, not for classic radiolabelled DD, due to the increased concentration of unlabelled dNTP which will be carried over to the DD reactions and which would reduce the incorporation of radiolabelled dATP to an unacceptable level.

As for all manipulations of RNA, maintain 'nuclease-free' laboratory conditions, wear clean gloves, and use aerosol-barrier, sterile, nuclease-free pipette tips. Set up all the reactions on ice.

3.6 Optimized fluorescent differential display with the fluoroDD system

The fluoroDD-PCRs must be carried out in duplicate to verify reproducibility. The reactions are more consistent when using a thermal cycler with a heated lid. Mineral oil overlays can be used in the event that a thermal cycler with a heated lid is unavailable, but the results may be less consistent.

Protocol 3

fluoroDD-PCR

Equipment and reagents

- 0.2 ml thin-walled MicroAmp (Perkin Elmer) PCR tubes (N801-0580) and caps (N801-0535)
- Thermal cycler: Perkin Elmer GeneAmp 9600
- First strand cDNAs (prepared according to *Protocol 2*)
- AmpliTaq®[a] DNA polymerase (5 U/µl) supplied with GeneAmp 10 × PCR buffer

- II[b] and 25 mM MgCl$_2$ stock solution (N808-0161, 250 U; or N808-0172, 1000 U) (Perkin Elmer)
- Genomyx HIEROGLYPH mRNA Profile Kit(s) (146081-85, Beckman Coulter, Inc.)
- Genomyx fluoroDD Kit (146086, Beckman Coulter, Inc.)
- Sterile nuclease-free H$_2$O (E476, Amresco; 9920, Ambion; or DEPC-treated H$_2$O[c])

Method

1. Thaw out the reagents[d] (first strand cDNAs, TMR 5′ end-labelled fluoroDD APs corresponding to the HIEROGLYPH APs used in the RT samples produced in *Protocol 2*, unlabelled HIEROGLYPH ARPs, dNTP mix, 10 × PCR buffer II, MgCl$_2$ stock solution, and AmpliTaq DNA polymerase) on ice, and keep them on ice. Mix each of the thawed reagents gently; do not vortex. Spin the tubes briefly to collect the contents in the bottom of the tubes.

2. For each cDNA subpopulation, prepare a fluoroDD-PCR core mix containing the appropriate RT sample and the matching TMR-AP in a volume sufficient for the number of reactions needed [No. rxns = (No. of 5′ ARPs) × (2 for duplicates) + 2]. Include all components in the core mix EXCEPT for the 5′ ARP. Aliquot the ARPs individually, 1.75 µl per PCR tube, into the appropriate PCR tubes. Add 8.25 µl of the appropriate core mix to each PCR tube, for a total volume of 10 µl per PCR tube. Mix samples well by gentle pipetting. Set up the positive and negative control reactions using the appropriate RT samples. Use only one ARP for the negative control DD-PCR.

3. The fluoroDD-PCR tubes will contain the following components in a final volume of 10 µl per reaction:

fluoroDD-PCR component	[Stock]	1 × vol. (10 µl)	[Final]
Sterile nuclease-free H$_2$O	–	1.95 µl	–
PCR buffer II (without MgCl$_2$)	10 ×	1.0 µl	1 ×
MgCl$_2$ stock solution	25 mM	1.5 µl	3.75 mM
dNTP mix (1:1:1:1)	250 µM each	2.0 µl	50 µM each
5′ ARP (exclude from core mix)	2 µM	1.75 µl	0.35 µM

3′ TMR-AP (fluorescent version of same 3′ AP in RT sample)	5 μM	0.7 μl	0.35 μM
RT sample (derived with the unlabelled form of same 3′ AP)	–	1.0 μl	–
AmpliTaq® enzyme	5 U/μl	0.1 μl	0.05 U/μl

4. Cap all of the tubes securely. Perform PCR in a thermal cycler with a heated lid (e.g. Perkin Elmer GeneAmp 9600). Other thermal cyclers may require protocol modification.

   (a) 95 °C for 2 min.

   (b) Four cycles: 92 °C for 15 sec, 50 °C for 30 sec, 72 °C for 2 min.

   (c) 30 cycles: 92 °C for 15 sec, 60 °C for 30 sec, 72 °C for 2 min.

   (d) 72 °C for 7 min.

   (e) Hold[e] at 4 °C.

5. Store the fluoroDD-PCR samples, wrapped in aluminium foil to protect them from light, at −20 °C or 4 °C.

[a] AmpliTaq® is a registered trademark of Hoffman-LaRoche, Inc. and Hoffman-LaRoche A.G.

[b] Be sure to use PCR buffer II, which does not contain Mg, and the separate $MgCl_2$ stock solution as directed, in order to obtain the correct final [Mg].

[c] See ref. 26.

[d] The TMR-AP and ARP stocks are provided at different concentrations. They are present at higher final reaction concentrations in the fluoroDD-PCR than in the classic radioactive HIEROGLYPH DD-PCR. The dNTP mix is present at a higher final reaction concentration in the fluoroDD-PCR also.

[e] It is strongly advised to avoid long thermal cycler hold periods at 4 °C (associated with loss of fluorescence signal due to dilution of samples from condensation within the cycler). If possible, schedule the cycling so that it is completed before the end of the day, and store the samples, wrapped completely in aluminium foil to exclude light, in a constant temperature −20 °C freezer or in a 4 °C refrigerator overnight, rather than leaving them in the cycler at 4 °C.

4 fluoroDD gel electrophoresis and imaging

4.1 fluoroDD gel electrophoresis

The GenomyxLRS™ System for fluorescent DD (230 V, 146518; 110 V, 146522; Beckman Coulter, Inc.) consists of the GenomyxLR DNA Sequencer for high resolution gel electrophoresis, and the GenomyxSC Fluorescent Imaging Scanner for data acquisition and analysis. Following fluoroDD-PCR, the TMR-labelled cDNA fragments are electrophoretically separated on a high resolution polyacrylamide gel (5.6% clear denaturing HR-1000 gel matrix) according to the detailed methods in the fluoroDD Kit manual.

Protocol 4

Preparation of high resolution fluoroDD gel

Equipment and reagents

- GenomyxSC Fluorescent Imaging Scanner with TMR and fluorescein (FL) filters (Beckman Coulter, Inc.)
- Genomyx Pour/Wash Tray (146124, Beckman Coulter, Inc.)
- Pair of low fluorescence GenomyxSC glass plates, 61 × 33 cm (146600, Beckman Coulter, Inc.)
- Pair of low fluorescence spacers, 61 cm long, 250 μm thick; and shark-tooth combs, 250 μm thick (48-well comb/spacer set, 146607; or 64-well comb/spacer set, 146608; Beckman Coulter, Inc.)
- 4 M NaOH
- Glass shield (solution, 146074; or towelettes, 146099; Beckman Coulter, Inc.)

- Genomyx Band Excision Workstation (146133, Beckman Coulter, Inc.)
- Genomyx Fluorescence Starter Kit (146136, Beckman Coulter, Inc.)
- Genomyx HR-1000 5.6% clear denaturing gel solution (146049, Beckman Coulter, Inc.)
- Sparkle or Dow Glass Plus glass cleaner (do not use ethanol or ammonia)
- 10% (w/v) ammonium persulfate (APS), freshly prepared with sterile nuclease-free H_2O
- TEMED
- Genomyx fluoroDD Kit manual (Beckman Coulter, Inc.)

Method

1. Follow the procedure for fluoroDD gel preparation given by the fluoroDD Kit manual. Be sure to add the grid position locator stickers to the proper points on the outer surface of the unnotched glass SC plate, before pouring the gel on the Pour/Wash Tray.

2. Gently mix 70 ml of 5.6% clear denaturing HR-1000 gel solution with 560 μl of 10% APS and 56 μl of TEMED, taking care not to introduce bubbles. Pour the gel and let it polymerize for at least 1 h.

3. Prior to loading the samples on the gel, take a flat field baseline scan of the gel using the GenomyxSC Scanner according to the fluoroDD Kit manual.

Protocol 5

fluoroDD gel electrophoresis

Equipment and reagents

- GenomyxLR DNA Sequencer (Beckman Coulter, Inc.)
- Polymerized HR-1000 5.6% clear denaturing gel (*Protocol 4*)
- 0.2 ml thin-walled MicroAmp (Perkin Elmer) PCR tubes (N801-0580)
- Thermal cycler: Perkin Elmer GeneAmp 9600

- 1 × and 0.5 × TBE buffer (146080, Beckman Coulter, Inc.)
- Genomyx fluoroDD Kit (146086, Beckman Coulter, Inc.)
- fluoroDD-PCR samples from *Protocol 3*
- Capillary gel loading pipette tips (AS-9152, Applied Scientific)

Protocol 5 continued

Method

1. Transfer 4.0 μl of each fluoroDD-PCR sample into new thin-walled 0.2 ml PCR tubes; add 1.5 μl of fluoroDD loading dye. Do NOT use the HIEROGLYPH sample loading dye, which will give high background fluorescence. Prepare the TMR-MW DNA standard markers from the fluoroDD Kit the same way (one tube containing 4.0 μl TMR-MW DNA standard markers plus 1.5 μl of fluoroDD loading dye for each gel).

2. Place the tubes, uncapped, in a thermal cycler and leave the lid of the cycler completely open. Heat the tubes at 95 °C for 2 min to denature and concentrate the samples so that they can be loaded entirely on the gel in a small volume (about 2.5–3 μl per lane). Quick-spin the tubes and keep them on ice. Save the remaining portion of the fluoroDD-PCR samples, securely capped and completely wrapped in aluminium foil to keep light out, at 4 °C or −20 °C.

3. Before loading the samples on the gel, be sure to perform a flat field scan of the gel using the GenomyxSC Scanner (*Protocol 4*). Then load the entire heat denatured sample per gel lane using the capillary gel loading pipette tips. Arrange the duplicates in adjacent lanes and place the samples generated with the same primer pairs in consecutive lanes in a consistent order.

4. The following fluoroDD electrophoresis conditions on the GenomyxLR DNA Sequencer are recommended for maximum resolution, optimal pattern determination, and cDNA fragment recovery. The 350 bp TMR-labelled DNA standard size marker will migrate near the bottom of the gel. Consult the fluoroDD Kit manual for further details.

GenomyxLR parameters	fluoroDD gel
Buffer in upper chamber	0.5 × TBE
Buffer in lower chamber	1 × TBE
Voltage	3000 V
Watts	100 W
Temperature	55 °C
Run time	5 h

4.2 fluoroDD gel analysis

After electrophoresis on a GenomyxLR DNA Sequencer, clean the outer surfaces of the fluoroDD gel plates with distilled water and then with glass cleaner; dry the plates with lint-free tissues. Place the fluoroDD gel, which is still between the two glass plates, in the GenomyxSC Fluorescent Imaging Scanner with the unnotched plate closest to the scanner camera. Collect the gel image data in accordance to the fluoroDD Kit and GenomyxSC manuals. Archive the data on compact disk. For best results, select and excise the candidate bands promptly.

4.3 Gel band selection and excision

Following the gel analysis on the GenomyxSC Fluorescent Imaging Scanner, take the notched plate off of the gel, leaving the gel on the full-length, unnotched glass plate. Dry the gel within the GenomyxLR Sequencer directly on the glass plate; this step has the advantages of eliminating cumbersome gel fixation, transfer to paper (decreasing the risk of tearing or stretching the gel), and vacuum drying. Then remove the urea from the gel according to the gel rinsing and drying process, which is detailed in the GenomyxLR and fluoroDD Kit manuals. The glass plate provides an ideal light-transparent support upon which to carry out cDNA band excision.

The Virtual Grid is a scanned image of the physical grid of the Band Excision Workstation. Apply the Virtual Grid template to the computerized gel image to locate the bands of interest. Note the position of these bands relative to the closest number and letter spot on the Virtual Grid; transfer the exact position of each band of interest to the physical grid with a fine point permanent ink marker. Be sure to record the ID number, approximate bp size, relative signal intensity, RNA sample, and AP-ARP primer pair for each candidate band.

It is best to select bands of high signal intensity that have significantly different expression levels for the control and experimental samples, and to consider only bands that are reproducible in duplicate lanes. For every band of interest, collect an equivalent area of gel from a corresponding control lane; this control 'band' or 'blank' should be located at the same migration distance from the gel origin as the target band. (If the cDNA is down-regulated in the experimental condition, the target band may be from the control sample lane while its 'blank' may be from the corresponding test sample lane.)

It is not necessary to cut out the entire band; only a small amount is needed. In fact, it is preferable to cut out only a portion of the band so as to minimize the amount of gel matrix and any extraneous contaminating cDNA bands. After cutting out the bands, the gel should be rescanned and the gel image overlaid with the Virtual Grid to check on the accuracy of band excision. If the band has not been accurately excised, carefully repeat the band cutting process and recheck the gel.

Consult the fluoroDD Kit manual for details on gel band excision and recovery. Excise each band with a new sterile scalpel blade (size 15, D2865-12, Baxter) and transfer it individually to a separate 1.5 ml sterile nuclease-free microcentrifuge tube. Then add 50 ml of TE to each tube, completely covering the gel slice, and heat the sealed tube in a 37°C water-bath for 1 h to allow the DNA to diffuse out of the gel matrix (do not boil the band). The TE eluent now contains a sufficient concentration of cDNA so that 2–4 μl is sufficient for most reamplification reactions. It is advisable to reamplify the cDNA fragments as soon as possible after excision. For long-term storage stability, it is best to subclone the fragments in a plasmid vector, as soon as possible after the reamplification. If the vector itself contains M13 reverse or T7 promoter sequences, then primers matching vector sequences flanking these sites should be used for sequencing the cDNA inserts.

5 cDNA fragment reamplification

The purpose of this step is to obtain sufficient amounts of cDNA for direct cycle sequencing and run-off transcription. The amplification is performed with full-length M13 reverse (–48) primer and full-length T7 promoter primer, to provide products that can undergo efficient primer annealing during cycle sequencing, and efficient promotion of RNA transcripts for RPA or Northern analyses.

Protocol 6

Reamplification of cDNA fragments from the fluoroDD gel

Equipment and reagents

- 0.2 ml thin-walled MicroAmp (Perkin Elmer) PCR tubes (N801-0580) and caps (N801-0535)
- Thermal cycler with heated lid: Perkin Elmer GeneAmp 9600
- Horizontal slab gel electrophoresis unit
- UV transilluminator or GenomyxSC Fluorescent Scanning Imager (Beckman Coulter)
- Genomyx Re-Amp Kit (146093, Beckman Coulter, Inc.) which includes optimized 5 × Re-Amp buffer, dNTP stock solution (2 mM each), and consensus primers: M13 reverse (–48) 24-mer primer (5′ AGCGGATAACAATTTCACACAGGA 3′), and T7 promoter 22-mer primer (5′ GTAATACGACTCACTATAGGGC 3′)
- 0.8% agarose gel in 1 × TAE buffer

- AmpliTaq® DNA polymerase (5 U/μl) supplied with GeneAmp 10 × PCR buffer II and 25 mM $MgCl_2$ stock solution (N808-0161, 250 U; or N808-0172, 1000 U) (Perkin Elmer)
- Sterile nuclease-free H_2O (E476, Amresco; 9920, Ambion; or DEPC-treated H_2O)
- Gel band eluent[a] from excised cDNA bands and corresponding 'blanks'
- 1 × TAE buffer: 0.04 M Tris–acetate, 0.001 M EDTA
- Gel loading dye from the HIEROGLYPH mRNA Profile kit (Beckman Coulter, Inc.)
- DNA standard molecular weight marker ladder (Gibco BRL Life Technologies)
- SYBR Green I (S-7563, Molecular Probes) or ethidium bromide[b]

Method

1. Follow the section on Gel Band Excision to recover a gel band containing a putative differentially expressed product. Collect the appropriate gel blank. If the target cDNA is 1.0 kb or longer, go directly to the SSCP gel protocol (*Protocol 7*).

2. Prepare a core mix using Genomyx 5 × Re-Amp buffer, the M13 reverse (–48) 24-mer sequencing primer (upstream), and the T7 promoter 22-mer sequencing primer (downstream). Aliquot 36.0 μl per reamplification tube.

Core mix component	[Stock]	1 × vol. (40 μl)	[Final]
Sterile nuclease-free H_2O	–	16.4 μl	–
Genomyx 5 × Re-Amp buffer	5 ×	8.0 μl	1 ×
dNTP mix (1:1:1:1)	2.0 mM each	3.2 μl	160 μM each
M13 reverse (–48) primer	2 μM	4.0 μl	0.2 μM
T7 promoter primer	2 μM	4.0 μl	0.2 μM
AmpliTaq enzyme	5 U/μl	0.4 μl	0.05 U/μl

Protocol 6 continued

3. Add individually to the appropriate tube:

   Gel band eluent (TE buffer)   4.0 μl.

4. Perform PCR in a Perkin Elmer GeneAmp 9600 thermal cycler with a heated lid:

   (a) 95 °C for 2 min.

   (b) Four cycles: 92 °C for 15 sec, 50 °C for 30 sec, 72 °C for 2 min.

   (c) 25 cycles: 92 °C for 15 sec, 60 °C for 30 sec, 72 °C for 2 min.

   (d) 72 °C for 7 min.

   (e) Hold at 4 °C.

5. Load 10 μl of reamplified product combined with 2 μl of loading dye on the agarose gel. Load the corresponding blank reaction sample in an adjacent lane. Load the DNA standard size marker on the gel.

6. Run the gel in 1 × TAE buffer at 80 V. Stain the gel with 50 ml of 1 × TAE buffer plus 5 μl SYBR Green 1 dye or with ethidium bromide. Visualize the DNA bands with long wavelength, low frequency, UVB light, or scan the agarose gel with the GenomyxSC Imager. If there are other bands in addition to the target band, it will be necessary to purify the appropriately sized band via SSCP (*Protocol 7*) before it is used in DNA sequencing, ribonuclease protection assays (RPA), or Northern blot screening.

7. Store the reamplified cDNA samples at −20 °C in a constant temperature freezer.

[a] In the event that an insufficient amount of reamplified cDNA is obtained, the reamplification can be scaled up and repeated using a greater amount of gel band eluent. To prevent the EDTA from inhibiting the PCR, do not use more than 5 μl of eluent in a 40 μl reaction volume unless the [EDTA] in the TE buffer used to elute the cDNA from the gel band is reduced to 0.5 mM or less. To prepare more reamplified cDNA, additional primary reamplifications using the eluent are preferable to secondary reamplifications using diluted aliquots of the primary reamplification. Secondary reamplifications tend to result in synthesis of non-specific products; in some cases the primary aliquots must be diluted 1:1 000 000 or more prior to the secondary reamplification.

[b] SYBR Green is approx. 25 × more sensitive than ethidium bromide for the detection of double-stranded nucleic acids. Use care in handling either of these potentially mutagenic stains.

6 Single strand conformation polymorphism (SSCP) gel

This gel is recommended for purification of high molecular weight cDNAs excised from DD gels, to remove extraneous species from the target cDNA band. It is advisable to reamplify the target cDNA fragments as soon as possible after excision from the SSCP gel. For long-term storage stability, it is best to subclone the post-SSCP cDNA fragments in a plasmid vector.

Protocol 7

SSCP gel purification of target cDNA bands

Equipment and reagents

- 0.2 ml thin-walled MicroAmp (Perkin Elmer) PCR tubes (N801-0580) and caps (N801-0535)
- Thermal cycler with heated lid: Perkin Elmer GeneAmp 9600
- Pair of 42 × 33 cm glass plates (146601, Beckman Coulter, Inc.)
- Pair of 42 cm long, 250 μm thick spacers; and shark-tooth combs, 250 μm thick (48-well comb/spacer set: 146605; or 64-well comb/spacer set: 146606; Beckman Coulter, Inc.)
- GenomyxLR DNA Sequencer (Beckman Coulter, Inc.)
- GenomyxLR Short Plate Adapter (146130, Beckman Coulter, Inc.)
- fluoroDD loading dye from the fluoroDD Kit (Beckman Coulter, Inc.)
- Bio-Max X-ray film (145100, Beckman Coulter, Inc.)
- GenomyxLR X-ray film cassette (146101, Beckman Coulter, Inc.)
- Sterile scalpel blades
- dNTP stock solution (250 μM each) from the HIEROGLYPH mRNA Profile Kit (Beckman Coulter, Inc.)
- TEMED

- Genomyx Re-Amp Kit (146093, Beckman Coulter, Inc.)
- AmpliTaq® DNA polymerase (5 U/μl) supplied with GeneAmp 10 × PCR buffer II and 25 mM $MgCl_2$ stock solution (N808-0161, 250 U; or N808-0172, 1000 U) (Perkin Elmer)
- $[\alpha\text{-}^{33}P]dATP$ (1000–3000 Ci/mmole specific activity) supplied at 10 mCi/ml (NEG612H, NEN Life Science Products; or AH9904, Amersham Life Science)
- Sterile nuclease-free H_2O (E476, Amresco; 9920, Ambion; or DEPC-treated H_2O)
- Gel band eluent from excised cDNA target bands and corresponding 'blanks'
- MDE non-denaturing gel, 2 × stock solution (50620, FMC)
- 20 × TTE buffer (1 litre): 216 g Tris base, 72 g taurine (Amersham), 4 g Na_2EDTA dissolved in sterile, nuclease-free water and adjusted to the final volume of 1 litre. Filter through a sterile 0.45 μm filter (Corning) to remove particulates.
- 10% (w/v) APS, freshly prepared with sterile, nuclease-free H_2O
- Sterile, nuclease-free TE buffer: 10 mM Tris–HCl pH 7.4, 1 mM EDTA

A. DD gel band reamplification pre-SSCP

1. Prepare the core mix and aliquot 18.0 μl per reamplification tube.

Core mix component	[Stock]	1 × vol. (20 μl)	[Final]
Sterile nuclease-free H_2O	–	7.95 μl	–
Genomyx 5 × Re-Amp buffer	5 ×	4.0 μl	1 ×
dNTP mix (1:1:1:1)	250 μM each	1.6 μl	20 μM each
M13 reverse (–48) primer	2 μM	2.0 μl	0.2 μM
T7 promoter primer	2 μM	2.0 μl	0.2 μM
AmpliTaq enzyme	5 U/μl	0.2 μl	0.05 U/μl
$[\alpha\text{-}^{33}P]dATP$	10 μCi/μl	0.25 μl	0.125 μCi/μl
(1000–3000 Ci/mmole specific activity)			

2. Add individually to the appropriate tubes for a 20.0 μl final reaction volume:

   Gel band or 'blank' eluent (TE buffer)   2.0 μl.

3. Reamplify the reactions in a thermal cycler with a heated lid:

   (a) 95 °C for 2 min.

   (b) Four cycles: 92 °C for 15 sec, 50 °C for 30 sec, 72 °C for 2 min.

   (c) 11 cycles: 92 °C for 15 sec, 60 °C for 30 sec, 72 °C for 2 min.

   (d) 72 °C for 7 min.

   (e) Hold at 4 °C.

B. Preparation of the SSCP gel

1. Clean and assemble 42 × 33 cm glass plates with 250 μm thick spacers. Filter the 20 × TTE buffer through a sterile 0.45 μM filter to remove particulates. Prepare the 0.5 × MDE gel.

   0.5 × MDE gel (70 ml)

2 × non-denaturing MDE stock	17.5 ml
Sterile, nuclease-free H$_2$O	50.4 ml
20 × TTE buffer	2.1 ml
10% (w/v) APS	280 μl
TEMED	28 μl

2. Degas the gel mix and pour the gel. Let it polymerize for at least 2.5 h.

3. The following SSCP electrophoresis conditions are recommended on the GenomyxLR:

Parameter	**SSCP gel**
Buffers in both chambers	1 × TTE
Voltage	1500 V
Watts	5 W
Temperature	25 °C
Run time	16 h

4. Denature 4.0 μl of reamplified cDNA sample with 1.5 μl of fluoroDD loading dye at 95 °C for 2 min: place tubes, uncapped, in the thermal cycler and leave the lid of the cycler completely open to allow evaporation of the aqueous fluid and concentration of the sample.

5. Chill the samples on ice. Load the target and corresponding control sample in adjacent gel lanes.

6. Insert two LR doorstops on the top edge of the GenomyxLR frame near the door hinge on the left side before closing the door and starting the run. This will keep the system temperature at approx. 35 °C (depending on ambient conditions) at the end of the run.

7. Dry the gel, wash once with distilled water for about 2 min, then dry thoroughly. Expose an X-ray film overnight on the dried SSCP gel.

Protocol 7 continued

8. Excise the major band[a] from each target lane, transfer it to a 0.2 ml thin-walled PCR tube, and cover the gel slice completely with 50 μl of TE. Incubate the tube at 37°C for 1 h.

9. Amplify an aliquot of the eluent using the reamplification protocol in part A, omitting the radioisotope, and cycling for 4 plus 25 cycles instead of 4 plus 11 cycles. The resulting sample can be used as a template for DNA sequencing. Store the samples at −20°C in a constant temperature freezer.

[a] The bands of interest are the more intense bands that appear in the reamplified target sample lanes and not in the corresponding controls. The minor bands are contaminating species or incomplete amplified products; these bands may also be observed in the corresponding control lane.

7 Direct cycle sequencing

If the reamplified DD-PCR product shows only one intense band of the correct size, purify the remaining reamplified product away from the excess primers and dNTPs with a spin column (A7170, Promega; or 28104, Qiagen) according to the manufacturer's instructions. Use 10 μl of eluate as the sequencing reaction template with the Thermo Sequenase dideoxyterminator cycle sequencing kit (US79750, Amersham). The M13 reverse (–48) primer will generate DNA sequence data from the coding (sense) strand which corresponds to the original mRNA transcript. The original AP used to produce the target band will generate the non-coding (antisense) DNA sequence.

8 Summary

Fluorescent DD provides further improvements in the reliability, speed, efficiency, and safety of the DD process. The protocols presented here offer advances in cDNA band reproducibility, resolution, analysis, recovery, and purification. The optimized kits and instrumentation systems that have been developed for fluorescent DD give researchers a robust and accessible approach for gene expression analysis.

Acknowledgements

I would like to thank all of my colleagues, both former and current, of the Genomyx Product Group at Beckman Coulter for their contributions to the development and success of the fluorescent differential display system, especially Dan Morrow, Peter Mansfield, and Frank Ruderman for their leadership and vision. I would like to express my appreciation in particular for the efforts of the Molecular Biology R&D team: Linda Reilly, Ph.D., Deepali Prabhavalkar, Kumar Kastury, Ph.D., Ching-Yi Wan, Ph.D., Cindy Leo; the Applications Support team: Dean Burgi, Ph.D., John Gripp, Tabassum Pittalwalla, Vikki Cerniglia, Don Kessler, Gary Osaka; the Engineering team: Hank Schwartz, Ph.D., Per Sjoman,

Jeff Land, Joe West, Reg Reyes, John Colby; and the BioKit Quality Control team: Lisa Milano, Elizabeth Pongo, John Gossett, and Jay Pluta. Our customers Gregory Shipley, Ph.D. and Harleen Ahuja, PhD. of The University of Texas at Houston Medical School developed the SSCP gel protocol, which we have adapted. I am grateful for the helpful discussions and feedback from our valued customers and research collaborators.

References

1. Liang, P. and Pardee, A. B. (1992). *Science*, **257**, 967.
2. Liang, P., Averbouk, L., Keyomarsi, K., Sager, R., and Pardee, A. B. (1992). *Cancer Res.*, **52**, 6966.
3. Guimaraes, M. J., Bazan, J. F., Zlotnik, A., Wiles, M. V., Grimaldi, J. C., Lee, F., *et al.* (1995). *Development*, **121**, 3335.
4. Hu, E., Liang, P., and Spiegelman, B. M. (1996). *J. Biol. Chem.*, **271**, 10697.
5. Babity, J. M., Newton, R. A., Guido, M. E., and Robertson, H. A. (1997). In *Methods in molecular biology: differential display methods and protocols* (ed. P. Liang and A. B. Pardee), p. 285. Humana Press, Totowa, NJ.
6. McCarthy, S. A., Samuels, M. L., Pritchard, C. A., Abraham, J. A., and McMahon, M. (1995). *Genes Dev.*, **9**, 1953.
7. Linskens, M. H. K., Feng, J., Andrews, W. H., Enlow, B. E., Saati, S. M., Tonkin, L. A., *et al.* (1995). *Nucleic Acids Res.*, **23**, 3244.
8. Duguid, J. R., Rohwer, R. G., and Seed, B. (1988). *Proc. Natl. Acad. Sci. USA*, **85**, 5738.
9. Sive, H. L. abd John, T. S. (1988). *Nucleic Acids Res.*, **16**, 10937.
10. Barr, F. G. and Emanuel, B. S. (1990). *Anal. Biochem.*, **186**, 369.
11. Herfort, M. R. and Garber, A. T. (1991). *BioTechniques*, **11**, 598.
12. Rubenstein, J. L. R., Brice, E. J., Ciaranello, R. D., Denney, D., Porteus, M. H., and Usdin, T. (1990). *Nucleic Acids Res.*, **18**, 4833.
13. Schweinfest, C. W., Henderson, K. W., Gu, J.-R., Kottaridis, S. D., Besbeas, S., Panotopoulou, E., *et al.* (1990). *Genet. Anal. Tech. Appl.*, **7**, 6.
14. Chou, S.-Y., Hannah, S. S., Lowe, K. E., Norman, A. W., and Henry, H. L. (1995). *Endocrinology*, **136**, 5520.
15. Wan, J. S., Sharp, S. J., Poirier, G. M.-C., Wagaman, P. C., Chambers, J., Pyati, J., *et al.* (1996). *Nature Biotech.*, **14**, 1685.
16. Welsh, J., Chada, K., Dalal, S. S., Cheng, R., Ralph, D., and McClelland, M. (1992). *Nucleic Acids Res.*, **20**, 4965.
17. McClelland, M., Honeycutt, R., Mathieu-Daude, F., Vogt, T., and Welsh, J. (1997). In *Methods in molecular biology: differential display methods and protocols* (ed. P. Liang and A. B. Pardee), p. 13. Humana Press, Totowa, NJ.
18. Lisitsyn, N. A. (1995). *Trends Genet.*, **11**, 303.
19. Hubank, M. and Schatz, D. G. (1994). *Nucleic Acids Res.*, **22**, 5640.
20. Ariazi, E. A. and Gould, M. N. (1996). *J. Biol. Chem.*, **271**, 29286.
21. Averboukh, L., Douglas, S. A., Zhao, S., Lowe, K., Maher, J., and Pardee, A. B. (1996). *BioTechniques*, **20**, 918.
22. Martin, K., Kwan, C.-P., and Sager, R. (1997). In *Methods in molecular biology: differential display methods and protocols* (ed. P. Liang and A. B. Pardee), p. 77. Humana Press, Totowa, NJ.
23. Bertioli, D. J., Schlichter, U. H. A., Adams, M. J., Burrows, P. R., Steinbiss, H.-H., and Antoniw, J. F. (1995). *Nucleic Acids Res.*, **23**, 4520.
24. Sheets, M. D., Ogg, S. C., and Wickens, M. P. (1990). *Nucleic Acids Res.*, **18**, 5799.
25. Chomczynski, P. and Sacchi, N. (1987). *Anal. Biochem.*, **162**, 156.
26. Sambrook, J., Fritsch, E. F., and Maniatis, T. (1989). Molecular Cloning, A Laboratory Manual, Second Edition, Cold Spring Harbor Laboratory Press.

Chapter 3

Practical aspects of the experimental design for differential display of transcripts obtained from complex tissues

ANDREW D. MEDHURST*, DAVID CHAMBERS[†],
JULIE GRAY*, JOHN B. DAVIS*, JULIAN A. SHALE[†],
IVOR MASON[†], PETER JENNER[‡], and
RICHARD A. NEWTON[§]

*Smithkline Beecham Pharmaceuticals, New Frontiers Science Park North, Third Avenue, Harlow, Essex CM19 5AW, U.K.

[†]Department of Developmental Neurobiology, UMDS, Guy's Campus, London SE1 9RT, U.K.

[†]Pharmacology Group, King's College, University of London, Chelsea Campus, Manresa Road, London SW3 6LX, U.K.

[§]Department of Physiology, Medical School, University Walk, Bristol BS8 1TD, U.K.

1 Introduction

Differential display is a valuable technique for identifying differentially expressed mRNAs and for cloning their cDNAs (1). Since its introduction in 1992, several technical improvements have increased the successful application of differential display (2–4). Many types of proteins (e.g. growth factors, chemokines, and pro-teases) have been identified using this technique, implicating their involvement in processes such as apoptosis, cell differentiation, and senescence (5). In addition, differential display constitutes a valuable tool for scientists interested in the identification of novel genes coding for proteins that may be suitable for pharma-cological intervention in diseases such as cerebral ischaemia (6) and cancer (7).

Despite numerous reports showing the success of differential display, and its advantages over other methods such as subtractive hybridization (e.g. the ability to simultaneously analyse multiple samples), widespread acceptance of the technique has not yet occurred. This has been partly due to technical short-comings attributed to factors such as inconsistent PCR materials, degraded RNA, or contaminating DNA (8), but mainly due to the risk of generating false positives. A false positive occurs when the expression pattern of a clone isolated from a differentially expressed cDNA band on a polyacrylamide gel cannot be reproduced, using independent techniques such as *in situ* hybridization, Northern blotting, or RT-PCR. Differential display is a multistep process where false positives can be generated and exacerbated at several stages, from the running of the experimental model and RNA extraction, through to cloning of candidate hits and confirmation of differential expression (*Figure 1*). Given this problem, confirmation of differential expression using one or more of these techniques, independent from the original differential display analysis, is an absolute requirement for claiming successful identification of a differentially expressed mRNA transcript.

The aim of this chapter is to discuss various practical aspects of experimental design when planning and carrying out a differential display analysis, in particular focusing on ways to reduce false positive generation, thereby helping to improve the chances of a successful differential display project. Good experimental planning at the outset, together with careful thought on whether the experimental system of interest is robust, reproducible, and suitable for complex differential display analysis, can all contribute to a reduction in time and resources wasted persuing false positive, differentially expressed mRNAs. In addition, careful experimental design further downstream, especially with cloning and confirmation techniques, should also help to reduce false positives

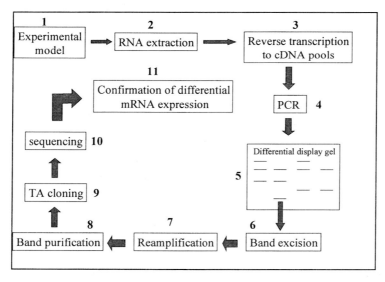

Figure 1 Potential sites in the differential display cascade for generation and exacerbation of false positives. For improved results using differential display:- (1) carefully plan and characterize experimental model before RNA extraction; (2) implement modified RNA procedures (or DNase treatment step) to reduce genomic DNA contamination; (3) run appropriate 'no reverse transcriptase' controls; (4) use Hieroglyph primers for increased specificity and reproducibility; (5) ensure differential expression is reproducible and occurs in duplicated lanes; (6) avoid excision of several or wrong bands by re-exposing gel after band excision; (7) reamplification may yield multiple products so select appropriate fragment; (8) use band purification step to help efficient ligation into cloning vector; (9) select at least eight positive clones; (10) sequence from both ends of cDNA to obtain maximum coding information; (11) if possible use independent RNA samples for confirmation using an independent technique.

and improve the success of differential display. These various points will be illustrated in this chapter with examples taken from several projects run in our laboratories. These examples will be drawn from four experimental paradigms in which we have used differential display:

- a mouse neuronal cell culture model of neurodegeneration
- a chick embryo model of the developing brain
- a rat model of neuropathic pain/diabetes
- a rat model of neurodegeneration

Each model has its own particular goals, complications, and experimental design specifics, while utilizing common protocols. We will discuss the 'suitability assessment' of each experimental system for differential display analysis, the preparation of samples for a differential display experiment, the generation and processing of putative bands, and finally the confirmation of these putative hits as true differentially expressed mRNAs using independent techniques. Several protocols we have utilized will be discussed throughout the chapter.

2 Suitability of experimental system for differential display

2.1 Characterization of experimental model and definition of clear goals?

At the outset it is important to consider how well the experimental system of interest is characterized, before undertaking a differential display analysis. A robust and reproducible model will increase the chances of successfully identifying differential gene expression. Ideally there should be a substantial and measurable biological difference between tissues or cells of similar background, and if possible, only one stimulus or condition should differ between samples. However, it is possible to investigate the effects of more than one stimulus as long as additional controls are included. Having some idea of the magnitude of biological response gives an indication of whether changes in gene expression are going to be easy to detect. It is much easier to study all-or-none patterns than more modest (e.g. twofold) changes in band intensity, on a differential display gel. A reproducible model is critical since an inherent variability of a biological effect, e.g. differences in lesion intensity, may complicate subsequent attempts to confirm differential expression. If the effect of drugs on mRNA expression *in vivo* are being investigated, it is important to establish that the correct blood level of drugs has been achieved in the appropriate target tissue, and for an appropriate time, before RNA is processed from the animals. There needs to be some evidence that the biological effect under investigation is mediated by changes in gene expression. For example, are transcriptionally active signalling cascades known to be activated in the model of interest, and if so, which gene families are they part of?

It is useful to establish which genes you are interested in before the experiment starts and therefore which time points may be most appropriate. If your interest is in immediate early genes or transcription factors, earlier time points maybe more relevant than later time points, when enzyme and receptor mRNA species may be of greater interest. Decide early on how many primer combinations to look at, and process some putative hits through to the confirmation step as early as possible, to validate the technique in your model before running too many primer sets. In our experience we found it better to screen a subset of primers and obtain early confirmation data rather than running a full screen of primers and confirming all putative hits together later on. If you decide to screen a small subset, then certain primers will be better than others, so select these accordingly (see Section 4.1.2). Be clear on how you would prioritise putative hits if you identify more than one category of hit. Be stringent in band selection by only selecting bands with large differences in intensity which can be replicated. Lack of rigour in candidate band selection may lead to an increased number of false positives and can also overload downstream steps unnecessarily. The strongest differences observed on the differential display gels are best placed as highest priority for processing through to confirmation of differential expression, although even these may only be small changes when subsequently quantified.

2.1.1 Introduction to four experimental paradigms in which we have used differential display

Our first experimental paradigm, the mouse HT4 neuronal cell line, is a well characterized immortalized cell line which has been used for the study of neuronal function (9). Oxidative stress has been implicated in the pathogenesis of neurodegenerative diseases and one potential mechanism for the generation of reactive oxygen species in conditions such as Parkinson's disease is via glutamate, which has been shown to influence the redox state of neuronal cells and increase their susceptibility to attack by reactive oxygen species (10). The HT4 cell line, or a subclone (HT22) with indistinguishable characteristics, has been shown to demonstrate several aspects of apoptotic cell death induced by glutamate, including protection by transcription and gene synthesis inhibitors (11, 12). The knowledge of the time course of rescue from apoptotic cell death by transcription inhibitors allowed the selection of an appropriate time point for collection of RNA (four hours), and also indicates that clear changes in gene expression should be identified in this model using differential display. Clones of this HT4 hippocampal cell line, which are sensitive or resistant to glutamate-mediated reactive oxygen species generation and cell death (11), were isolated for use in differential display. We utilized differential display to identify genes involved in glutamate-induced apoptosis, and also genes that confer resistance to glutamate-induced toxicity. RNA was only extracted when parallel cultured cells demonstrated the correct phenotypic response to glutamate application.

The developing vertebrate hindbrain consists of cell lineage-restricted segmental units known as rhombomeres, which constitute domains of specific gene expression (13). The segmental pattern of rhombomeres is responsible for the ordered organization of motor and reticular neurons. Several genes, including Hoxa1 and Hoxb1, have already been identified to exhibit rhombomere-specific patterns (14) but to date, no molecular markers have been identified which solely distinguish rhombomere 1 from the surrounding regions of the neural tube. Therefore, since rhombomere-specific developmental genes do exist, there was good rationale for our second differential display paradigm. Hence, the aim of this project was to identify rhombomere 1-specific genes in the developing chick brain, as well as identifying other rhombomere-specific gene expression patterns that may increase the understanding of brain segmentation in development.

For our third experimental system we utilized a rat model of chronic neuropathic pain, based on partial ligation of the sciatic nerve (15), to identify genes associated with the pain following nerve injury. Ligation results in varying degrees of deformity of the foot and postural changes as well as sensitivity to external stimuli (16). Previous studies have demonstrated changes in gene expression in such models. For example vasointestinal polypeptide (VIP) mRNA is known to be up-regulated after nerve injury (17). Since diabetes has been linked to painful neuropathy (18), our differential display analysis was carried out on RNA obtained from Zucker Diabetic Fatty (ZDF) rats as well as lean non-diabetic rats, both exposed to partial sciatic nerve ligation injury. Our goal was to identify gene changes associated with pain induced by nerve ligation, but also those changes additionally associated with painful neuropathy during diabetes. RNA was only isolated from those animals that demonstrated hyperalgesia 14 days following partial sciatic nerve ligation.

The 6-hydroxydopamine (6-OHDA)-lesioned rat is widely used as a model for Parkinson's disease. Unilateral destruction of the dopaminergic nigro-striatal tract results in a permanent imbalance in striatal output pathways between the two hemispheres, which produces a characteristic rotation response following administration of dopamine agonists (19). In this model, changes occur in gene expression in the direct and indirect GABA output pathways from the striatum. For example, preprotachykinin mRNA is down-regulated, whilst preproenkephalin mRNA is up-regulated after lesioning (20, 21). These changes in expression can be normalized by the use of specific D1 and D2 dopamine receptor agonists (22). The aim of this project was to identify changes in gene expression within the striatum induced by 6-OHDA lesioning, which may further our understanding of the neuronal basis of Parkinson's disease. Consequently, RNA was extracted from the tissues of 6-OHDA-lesioned animals showing the characteristic behavioural response to dopamine agonists.

2.2 Determination of appropriate controls and sample replication

One of the advantages of differential display is the capacity to use multiple lanes to study several RNA populations simultaneously from small amounts of starting material. This allows the incorporation of many controls to direct selection of priority band changes. Proper controls will help the selection of potential biologically important changes and also minimize the chances of selecting spurious bands. Too few controls increase the possibility of selecting false positives, whilst too many controls could make the differential display analysis too complex and gel interpretation more difficult. A good example of a control would be to use ipsilateral and contralateral tissues from the same animal if the experimental treatment is unilateral.

Due to the complex multistage nature of differential display analysis as we outlined earlier (*Figure 1*), both replication and reproducibility are important considerations in this technique. It is essential that only reproducible changes between replicates are considered for further analysis. This 'reproducibility check' can be incorporated at several stages in the differential display process, depending on the experimental model of interest. If there is enough material available, replication at the RNA preparation step is possibly the best option. For each experimental condition, preparation of two independent RNA pools works well, especially when the samples are processed in parallel to minimize variation between populations. Another alternative is to use RNA from individual animals as your replication step. Analysis of expression in individual animals gives a better idea of variability, but often the small quantity of the tissue and RNA of interest will necessitate sample pooling (e.g. individual rhombomeres from chick embryos). If you decide to pool RNA rather than tissues, try to prepare the RNA from each individual animal so you can always go back to study variability between animals later, and pool as many RNA samples as possible, in order to limit the effect of spurious samples.

Duplicating the reverse transcription stage is another option, but this may be difficult when RNA is limited. Using TaqMan™ real time quantitative PCR analysis (23) we have observed that preparing two reverse transcription reactions from the same RNA sample results in almost superimposable PCR signals. Alternatively you may decide to run duplicate PCR lanes. Again this is usually very reproducible when measured quantitatively, and may be the only option if there is very little RNA available. However, this could be considered as false replication since you will only be replicating PCR, and hence any errors it produces. Finally, if RNA availability is not a problem you could introduce the replication step at the confirmation stage. Once a putative hit has been identified by differential display, a Northern blot could be carried out for example, on RNA prepared from several batches of cells. Alternatively, if a putative hit was generated from two animals, confirmation could be checked in other animals independent from the original differential display analysis. With the advent of 96-well plate techniques such as TaqMan™ real time quantitative PCR, it is now possible to include several time points, individual animals, or other cDNAs in a 'differential display confirmation' TaqMan™ plate, where numerous samples can be analysed simultaneously and quantitatively.

In the HT4 model we ran duplicate PCR reactions from each cDNA, as well as replicating the confirmation step (see Section 6). For controls in the HT4 cell project, in addition to wild-type (WT) HT4 cells and WT cells treated with glutamate (WTG), we included a second set of samples from resistant clones (RC) of the HT4 cell line, that do not die in response to glutamate, and the same RC also treated with glutamate (RCG). In this way we hoped to exclude hits that were due to non-specific effects of glutamate. The differential display analysis therefore incorporated four conditions: WT, WTG, RC, and RCG (total of eight lanes per primer combination).

Duplicate RNA samples, prepared from independent pools of tissue were used in the chick embryo project. Total RNA was extracted from 327 individual neural tube regions in stage 10/11 chick embryos and subdivided into two RNA pools. Since the aim of the project was to investigate region-specific genes, especially those unique to rhombomere 1, RNA was prepared from hindbrain (HB, consisting of rhombomeres 2–7) and midbrain (MB), as well as from rhombomere 1 (R1), which had been cleared of other adhering tissues such as notocord and mesoderm. Thus the differential display analysis comparison was between RNA from R1, HB, and MB (total of six lanes per primer combination).

We also utilized duplication of RNA pools in the rat neuropathic pain/diabetes model. RNA was prepared from left and right dorsal root ganglia (DRG) isolated from the ipsilateral and contralateral side to the nerve ligation of hyperalgesic lean (LN) and diabetic (ZDF) animals. Tissues were processed individually and only pooled after RNA extraction. RNA from five rats and six rats constituted pool A and pool B respectively. In addition, sham lean and sham diabetic animals were included, but only one pool of RNA was generated since fewer animals were available. This gave a 12 lane differential display analysis, i.e. two independent RNA pools (A and B) of ligated ZDF (L and R), ligated LN (L and R), as well as one RNA pool from sham ZDF (L and R) and sham LN rats (L and R). The aim of this project was to identify any changes in gene expression during pain associated with nerve ligation or nerve ligation-plus-diabetes.

In the 6-OHDA lesion model, duplicate PCR was run from each cDNA template, with additional replication being achieved by analysing two samples from independent rats in parallel (R1 and R2) on the differential display gels. Since this model involves the unilateral effects of 6-OHDA injection, the striatum from the contralateral hemisphere (CS) acted as an internal control for the striatum ipsilateral (IS) to the lesion in each animal. The differential display analysis was therefore a comparison of ISR1, CSR1, ISR2, and CSR2 (eight lanes per primer combination).

2.3 Selection of method for confirmation of differential expression

Careful selection of an experimental system where there is a reproducible source of samples for the investigations required to confirm the differential regulation pattern, as well as the subsequent cloning and characterization, is critical. It is relatively easy to obtain differences in band patterns on differential display gels in most models, but the downstream processing and verification of differential expression can be complex.

It is very important to determine at the outset how putative hits identified in differential display will be confirmed, as this will dictate how much material will be required. Often the model will dictate which confirmation technique is most appropriate. Confirmation of hits using independent techniques is crucial, given the frequent identification of false positives. Downstream verification of any putative hits is required using techniques such as Northern blotting, *in situ* hybridization, RT-PCR, TaqMan™ quantitative PCR, or RNase protection assays. *In situ* hybridization can be time-consuming for confirming many hits and the sensitivity can be too low for some putative bands. Northern blot analysis requires much larger amounts of RNA, and is therefore not very sensitive. RNase protection is more sensitive, but still requires substantial amounts of RNA and depends on careful optimization for each target of interest, whilst TaqMan™ real time PCR is quantitative, sensitive, and suitable for several hits, although it can be expensive.

In the HT4 cell project we chose to confirm putative hits first by Northern blotting, but later by TaqMan™ quantitative PCR using RNA from three independent batches of cells. Due to the very limited tissue (and therefore RNA) quantities obtained from the chick embryos, putative differentially expressed hits were screened using whole mount *in situ* hybridization in stage 10/11 chick embryos. In the rat neuropathic pain/diabetes model, since some putative hits appeared as 'all-or-none', relative RT-PCR could be used for initial confirmation prior to a more quantitative determination using TaqMan™ analysis. Finally, TaqMan™ real time PCR analysis was used on RNA from two rats to confirm hits generated from the 6-OHDA lesioning model.

3 Preparation of samples for differential display

3.1 Tissue collection and homogenization

Tissue dissection should be performed using sterile, RNase-free surgical instruments. If possible it is preferable to perform subdissections of tissue (e.g. specific brain regions from whole brain) before freezing the tissue in liquid nitrogen, since it is more difficult to obtain good quality RNA if the tissue requires partial thawing for subsequent subregional dissection. Perform these dissections on ice whenever possible, and try to minimize dilution effect by dissecting free very specific tissues of interest (e.g. dentate gyrus) rather than whole organs (such as brain). We found it most practical to freeze the tissue in liquid nitrogen, keep it on dry ice during transportation, and store at $-80\,°C$ as soon as possible, rather than to process RNA from freshly dissected tissue, especially when preparing numerous samples. In fact we often found that homogenizing frozen tissue resulted in better RNA quality than homogenizing fresh tissue. In addition, we found it easier to wrap liquid nitrogen-frozen tissue in labelled aluminium foil for storage at $-80\,°C$, rather than in cryo tubes, which make tissue retrieval more difficult. A small amount of antifoam agent often helps when processing protein-rich samples such as brain material. We regularly store 1 ml aliquots of TRIzol homogenates (at $-80\,°C$) for subsequent RNA extraction (see *Protocol 1*).

3.2 Total RNA extraction

There are numerous methods for total RNA extraction but in our laboratories we have successfully used the TRIzol method (Gibco BRL) for differential display. We have modified the original protocol to help decrease any genomic DNA contamination, and therefore reduce the necessity for DNase treatment of total RNA which is not always practical (see Section 3.3). The only difficulty we encountered was when insufficient TRIzol reagent was used per square centimetre of culture dish or per milligram of tissue when some DNA contamination occurred. Passing the homogenate carefully through a 19 gauge needle attached to a syringe a few times also helps shear genomic DNA. Regardless of which RNA protocol is used, standard precautions for avoiding RNA degradation should be followed, including wearing gloves and using RNase-free solutions and glassware. Whenever possible we harvested RNA from all the different conditions at the same time to reduce variability.

Protocol 1

Modified TRIzol reagent protocol for total RNA extraction

Equipment and reagents

- RNase-free tubes (1.5 ml)
- Homogenizer (e.g. Polytron)
- Refrigerated bench-top centrifuge (e.g. Eppendorf)
- TRIzol reagent (Gibco BRL)
- Chloroform
- Isopropanol
- 75% ethanol
- RNase-free sterile water

Method

1. Homogenize tissue samples in 1 ml TRIzol reagent per 50–100 mg tissue. For cells grown in monolayer, lyse cells directly in a culture dish or flask by adding 1 ml TRIzol reagent per 3.5 cm diameter dish. If many samples are being processed simultaneously, TRIzol aliquots can be stored at $-80\,^{\circ}\mathrm{C}$ for several months, but ensure that the homogenized samples are left at room temperature for 5–10 min before freezing to allow complete dissociation of nucleoprotein complexes. In addition, allow frozen aliquots to thaw to room temperature before proceeding to step 2.

2. Add 0.2 ml of chloroform per 1 ml TRIzol, cap tubes securely, and shake vigorously for 15 sec. After incubation of samples at room temperature for 2–3 min, centrifuge samples (12 000 g) at $4\,^{\circ}\mathrm{C}$ for 15–20 min. After centrifugation, carefully remove the colourless upper aqueous phase (containing RNA) from the lower red phenol: chloroform phase and interface, and transfer to a fresh tube.

3. To obtain purer RNA we have utilized one of two additional steps.

   (a) *Either* perform a double extraction by adding more TRIzol (0.5 ml) and repeating step 2.

    (b) *Or* centrifuge the aqueous phase for 2–3 min at 4°C and transfer the aqueous phase to another fresh tube without disturbing any residual interface or phenol:chloroform remaining. This step may not be desirable if RNA yields are expected to be low.

4. To precipitate the RNA from the aqueous phase, add 0.5 ml isopropanol per 1 ml TRIzol and mix briefly by inverting capped tubes three or four times. Incubate samples at room temperature for 10 min and then centrifuge (12 000 g) for 10 min at 4°C. After centrifugation a small gel-like pellet should be visible at the bottom of the tube.

5. Carefully remove the supernatant without disturbing the pellet and wash by adding 1 ml of 75% ethanol per 1 ml TRIzol used for homogenization. Mix the sample by vortexing, and centrifuge (no more than 7500 g) for 5 min at 4°C. We often incorporate an additional washing step here by repeating step 5 which can help to improve RNA purity. This may not be desirable if RNA yields are expected to be very low.

6. Briefly air dry or vacuum dry the RNA pellet and resuspend the pellet in 20–50 μl of RNase-free sterile water. Incubate sample at 55°C for 10 min if pellet is not dissolving properly.

7. Check RNA quantity using A_{260} ($\times$ 40 $\times$ dilution) measurement, and confirm RNA integrity by running samples on 1–1.5% agarose gel with 1 $\times$ TBE buffer (5 μl sample and 5 μl of 6 $\times$ loading dye).

8. Aliquot RNA samples to required concentrations and quantities for differential display to avoid freeze–thawing.

3.3 DNase treatment

Depending on the quality and quantity of total RNA preparations it may be necessary to DNase treat samples before running differential display, particularly in instances where duplicates are from the same RNA source since any spurious DNA contamination will be present in both samples. However, if RNA quantity is limiting, too much RNA may be lost during downstream processing such as DNase treatment. We rarely found spurious bands appearing in differential display lanes from 'no reverse transcription' controls when using the modified RNA preparation described in *Protocol 1*. We have not routinely DNase treated our RNA samples, although we have done occasionally when there was a clear problem of bands appearing in no reverse transcription lanes. However, on further analysis, these spurious bands were attributed to contamination of a PCR reagent. We also have some experience with TaqMan™ real time quantitative PCR analysis on cDNA samples prepared from RNA treated with different DNase enzymes and buffers. In particular, when a DNase reaction was run at 37°C for 60 min with a high concentration of $MgCl_2$ in the buffer, Mg^{2+}-dependent hydrolysis of RNA appeared to occur, as determined by a dramatic decrease in quantitative TaqMan™ PCR signal. This occurred despite claims that this protocol should not affect RNA when analysed by gel electrophoresis. Our best results were obtained using DNase protocols requiring a short incubation at room temperature, followed by termination of reaction with EDTA (e.g. DNase I amplification grade, Gibco BRL). In addition, avoiding any subsequent phenol:chloroform extraction helps to prevent further loss of RNA sample.

3.4 Reverse transcription

We have successfully utilized the reverse transcription protocol based on the one supplied with the Genomyx Hieroglyph primers (Beckman). We have found that the use of SuperScript II reverse transcriptase (Gibco BRL) as in *Protocol 2* gives sharper differential display band patterns and longer products than other commercial enzymes. We generally only run the 'no reverse transcription' controls for the first few primer combinations, until we have established the absence of spurious bands attributed to DNA contamination.

Protocol 2

First strand cDNA synthesis

Equipment and reagents

- Sterile PCR tubes or plates
- Thermal cycler (e.g. PTC-200, MJ Research)
- SuperScript II RNase H⁻ reverse transcriptase (200 U/μl) supplied with 5 × first strand SuperScript II buffer (250 mM Tris–HCl pH 8.3, 375 mM KCl, 15 mM MgCl₂) and 100 mM DTT (Gibco BRL)
- 20 U/μl RNasin (Promega)

- 250 μM dNTP mix (Genomyx Hieroglyph kit, Beckman)
- 2 μM anchored primer (Genomyx Hieroglyph kit, Beckman)
- RNase-free sterile water
- 200 ng/μl total RNA template

Method

1. On ice, add RNA template and 2 μl anchor primer into PCR microtube or plate (depending on how many reverse transcription reactions you are preparing).

2. Incubate at 65 °C for 5 min and then cool on ice for at least 2 min.

3. Prepare on ice a core master mix of the following reagents in sufficient volume for the number of reactions required (+10%). Also prepare a separate master mix replacing the reverse transcriptase with water for use as a control. All volumes given are per reaction.

   - 4 μl SuperScript II buffer
   - 2 μl DTT
   - 2 μl dNTP mix
   - 1 μl RNasin
   - RNase-free sterile water to 16 μl
   - 1 μl SuperScript II reverse transcriptase

Protocol 2 continued

4. Add 16 μl of master mix into appropriate tubes or wells for a final volume of 20 μl per reaction. Mix well and quickly spin tubes or plate.

5. Perform first strand synthesis reactions on thermal cycler block using the following parameters: 25 °C for 10 min, 42 °C for 60 min, 70 °C for 15 min, and 4 °C holding temperature.

6. Store generated cDNA products for differential display at −20 °C.

4 Generation of putative hits using differential display

4.1 Choice of equipment and reagents

4.1.1 GenomyxLR sequencing tanks and HR-1000 denaturing gels

We originally used conventional 45 cm long sequencing gels for differential display, but these gave limited resolution of only small fragments, and so allowed the isolation of small ESTs (< 400 nt) only. Gel 'smiling' often occurred, unless gels were pre-run for substantial periods. In addition, buffer loss sometimes occurred resulting in overheating of the gel plates. We then used the GenomyxLR gel electrophoresis system which improved results dramatically. Since the GenomyxLR system also contains a gel drying system and electrical power pack, it is a wise purchase when starting out on a first differential display project. The 61 cm long gels allow for the resolution of more bands, and clearer resolution of large fragments. We have never experienced gel smiling with this system, probably due to the carefully controlled heating system. Overnight gel runs are not a problem, allowing full resolution of products of 2–2.5 kb when 4.5% denaturing gels are used. The increased resolution enhanced the ability to isolate large cDNAs which have a greater possibility of containing sequence within the coding region. Using shark-tooth combs produced no problems with sample leakage across wells and allowed the running of 64 lanes instead of the standard 48 lanes. By drying gels directly onto the glass plate, the resolution on autoradiograms was also improved. Together with the Hieroglyph primers described below, this equipment has greatly improved our assays.

4.1.2 Hieroglyph primers

We have previously utilized the classical differential display primers as described by Liang and Pardee (1), but have found that the Hieroglyph primers gave more reproducible results.

The reproducibility of mRNA differential display is critically dependent on the specificity of the primers and the stringency of the PCR conditions. The use of long anchored and arbitrary primers allows high temperature annealing after initial cycles of low stringency annealing. Therefore, randomly primed products generated in the initial cycles are accurately amplified during subsequent cycles, giving improved reproducibility and sensitivity. The anchored primers we used incorporate two nucleotides immediately upstream of an oligo(dT)$_{12}$ region. There are 12 possible two-base anchored primers which each target one of the 12 possible combinations of the two bases immediately upstream of the poly(A) tail, omitting T as the penultimate base. High stringency annealing is facilitated by the incorporation of a 17 nucleotide partial T7 promoter sequence at the 5′ end of the primer, a sequence which also allows the generation of antisense transcripts for subsequent confirmation experiments. The addition of unrelated sequences at the 5′ end of the primer does not generally alter the annealing of the sequence-specific portion of the primer (24). The arbitrary primers are 26 nucleotides in length with the terminal 10 bases at the 3′ end forming the core annealing sequence and a partial M13 reverse sequence at the 5′ end. All of the clones so far examined were generated by the combination of arbitrary and anchored primer, although in some cases the anchor primer did not hybridize to the poly(A) tail but to sequences within the coding region.

Our screening strategy initially utilized the various arbitrary primers in combination with each of the four anchored primers that possess a 'G' in the penultimate position. The preferential use of these four anchored primers is due to the fact that approximately 60% of all mRNA species in vertebrates have a CA dinucleotide directly upstream of the poly(A) tail (25). Thus if you are planning to screen only a subpopulation of primer combinations it is sensible to select these primers first. In addition, some primers continuously gave poor band patterns. Also, T12CT, T12AT, and T12GT often gave more smeared banding, perhaps due to a lack of specificity when annealing to the poly(A)$^+$ segment of mRNAs, and thereby generating artefacts. We have noticed that there was some redundancy in the displays with different anchored primers, i.e. we identified certain cDNA fragments more than once and with different primer combinations. An assessment of the number of different cDNA species visualized, based on the number of bands per gel lane could therefore lead to an overestimation of mRNA coverage. However, this is somewhat offset by the occupation of the same band positions by more than one cDNA species.

4.2 Preliminary experiments

4.2.1 RNA titration for differential display

One of the many advantages of differential display is the ability to analyse differences in gene expression using relatively small starting amounts of total RNA. However, to obtain maximal coverage of the RNA pools of interest it is necessary to perform a high number of different primer combinations, which lead to increasing demands on the total amount of RNA required. In cases where available RNA quantities are a limiting factor, it is an advantage to perform an RNA titration experiment prior to the differential display study, to determine the absolute minimum amount of material required for all planned primer combinations to be analysed. In our experience, such experiments have revealed that a single reverse transcription reaction (sufficient for ten PCR reactions) can be performed from as little as 50 ng total RNA without loss of the characteristic differential display banding pattern. Indeed, as little as 20 ng of input RNA per 20 μl RT reaction has been reported to be sufficient for differential display (26). However, it is often wise to use more than the absolute minimum amount to compensate for any variations in OD readings for example.

4.2.2 Targeted display/targeted upstream primers

If a particular gene is already known to be differentially regulated in the model of interest, a good preliminary experiment is to target this gene specifically to check that the differential display system can identify this change. In addition to utilizing upstream primers whose 3′ core annealing sequences were of an arbitrary nature, we designed 26-mer upstream primers which had 14 nucleotides at the 3′ end that matched those of a known gene. When the sequence of the 3′ untranslated region (UTR) is available, the appropriate anchor primer can be selected and the size of the product predicted. By targeting a gene whose mRNA expression is known to be altered by the experimental paradigm adopted, a validation of the experimental methods can therefore be obtained.

Such a positive control was applied to the rat model of neuropathic pain in our laboratory by targeting rat VIP mRNA, a message that is known to be altered following sciatic nerve injury (17). A degenerate glucagon polypeptide family upstream primer was designed whose 3′ terminal region was homologous to a region of VIP mRNA approximately 1 kb upstream of the poly(A) tail. This upstream primer was used in our differential display protocol (*Protocol 3*) in combination with a T7(dT)$_{12}$CG anchor (since GC are the two nucleotides adjacent to the poly(A) tail), and resulted in the identification of a differentially expressed band of the expected size (*Figure 2*). Since in differential display the initial primer annealing is carried out at low stringency, many transcripts unrelated to VIP were also amplified under these conditions. The targeted primers therefore also function as additional arbitrary primers for screening the mRNA population. Differential display utilizing upstream primers that carry a sequence motif have previously been used to target particular families of genes (27). In such studies, it is essential that a highly conserved region is chosen because for long primers, at least eight nucleotides at the 3′ end need to be 100% homologous to the target sequence (28).

4.3 Differential display run

Most of our differential display analyses (especially overnight gel runs) were carried out on 4.5% polyacrylamide gels with the GenomyxLR sequencing apparatus, although for short gel runs, 6% gels yielded clearer band patterns. By using 64-well shark-tooth combs we maximized sample numbers on each gel without sacrificing clarity of band patterns. [α-^{33}P]dATP gave sharper signals than [α-^{35}S]dATP and did not work out more expensive since 20% less volume than [α-^{35}S]dATP was used per reaction. It is helpful to run size markers (e.g. generated with M13 DNA from the Sequenase kit, Amersham) on the first few gel runs, to estimate the fragment sizes of differential display products and to

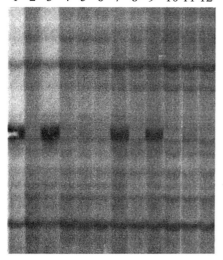

Figure 2 Section of a differential display gel obtained using VIP-targeted arbitrary primer and a dT$_{12}$CG anchor primer. The differentially expressed band was of the predicted size (approx. 1.1 kb) and occurred in lanes 1, 3, 7, and 9 using RNA from DRG of nerve-injured animals. No signal was seen with reactions using RNA from contralateral DRG of nerve-ligated animals (lanes 2, 4, 8, and 10), nor with ipsilateral (lanes 5 and 11) or contralateral (lanes 6 and 12) DRG of sham-operated rats. The region of gel fragment excision for the differential band is visible in lane 1 and the subsequent reamplification, cloning, and sequencing confirmed the band as VIP.

judge how far to run the gels. We tended to run one anchor primer per gel against several arbitrary primers but this is obviously dependent on the way each researcher designs their PCR experiments.

Protocol 3

Differential display PCR using Hieroglyph primers and GenomyxLR sequencing apparatus

Equipment and reagents

- GenomyxLR sequencing apparatus (Beckman)
- HR-1000 high resolution 4.5% differential display gel (Beckman)
- PCR plates or tubes
- Thermal cycler (e.g. PTC-200, MJ Research)
- 5′ arbitrary and 3′ anchored primers (Genomyx Hieroglyph kits, Beckman)
- 10 × PCR buffer II: 100 mM Tris–HCl pH 8.3, 500 mM KCl (Perkin Elmer)

- 25 mM MgCl$_2$ (Perkin Elmer)
- RNase-free sterile water
- 250 μM dNTP mix (Genomyx Hieroglyph kit, Beckman)
- cDNA derived from appropriate 3′ anchor primer
- 5 U/μl AmpliTaq polymerase (Perkin Elmer)
- 10 μCi/μl [α-^{33}P]dATP (Easitide, NEN)

Method

1. On ice, add 2 μl 3' anchor primer and 2 μl cDNA to appropriate wells of a PCR plate (banding patterns are less reproducible if not prepared on ice).

2. Set up core master mix in sufficient volume for the number of PCR reactions (+10%) as follows (volume per reaction):
   - 2 μl of 10 × PCR buffer
   - 1.2 μl MgCl$_2$
   - 1.6 μl dNTP mix
   - 2 μl 5' arbitrary primer
   - 8.75 μl sterile water
   - 0.25 μl [α-^{33}P]dATP
   - 0.2 μl AmpliTaq enzyme

3. On ice, add 16 μl master mix to each well mixing carefully. Perform PCR in a thermal cycler with heated lid or with a drop of mineral oil (Sigma) covering samples, using the following parameters:
   (a) 95 °C for 2 min.
   (b) Four cycles of 92 °C for 15 sec, 46 °C for 30 sec, 72 °C for 2 min.
   (c) 25 cycles of 92 °C for 15 sec, 60 °C for 30 sec, 72 °C for 2 min.
   (d) 72 °C for 7 min.
   (e) 4 °C holding temperature.

4. Mix 7 μl PCR sample with 4 μl of sample loading dye and denature in thermal cycler for 2 min at 95 °C. Quick spin the samples and then keep on ice. Load 3 μl of sample per lane, arranging duplicates in adjacent lanes and comparative experimental samples as appropriate.

5. (a) Perform a short screening gel (3 h run, 50 °C, 3000 V, 100 W).
   (b) For resolution and excision of bands larger than 700 nucleotides, perform an overnight run (50 °C, 1250 V, 100 W). Pour 4.5% gel about 90 min before use.

6. After electrophoresis run is complete, carefully separate plates and place the lower plate containing the gel back into the GenomyxLR tank to dry down the gel. Carry out three to four washes with water of the dried gel to remove excess urea that appears as crystals during drying, replacing the plate back into the dryer between each wash.

7. Once the gel is dry and free from excess urea, lay down against Biomax film (Beckman) overnight.

8. Select putative differentially displayed bands by eye and try to choose the best differences for processing first.

5 Processing of putative differentially displayed hits

5.1 Candidate band excision

Once a putative differentially expressed cDNA has been identified, further analysis is required to determine both the identity of the candidate cDNA and also to confirm the expression pattern. The band of interest is excised from the gel and reamplified at the exclusion of neighbouring bands that are not differentially expressed. This is a particular step in the differential display process that has clearly contributed to the false positive rate, i.e. isolation of the wrong band from the gel, or isolation of more than the candidate band. Using the GenomyxLR system we were routinely able to successfully excise appropriate gel segments whilst the dried gel was still attached to the glass plate, therefore making it unnecessary to lift the gel onto paper (*Protocol 4*). In addition, no boiling of the gel slice is required, but it is placed immediately into a PCR tube. Moreover, unlike other reamplification protocols, it is not necessary to remove the gel slice from the final amplification reaction. Despite the advent of better gel systems with improved resolution, it is still possible to inadvertently excise more than one band. A simple but effective check for this is to re-expose the gel after band excision to confirm precise removal of the gel slice of interest (*Figure 3*). To further increase the chances of recovering only the band of interest we implemented several simple steps. The differential display gels should be run for as long as possible, to maximize the resolution of the candidate band. Using the GenomxyLR gel apparatus we have run 4.5% gels for up to 18 hours to resolve candidate bands over 1 kb. Once the candidate band has been marked, the portion of gel to be removed is rehydrated using sterile water to allow the gel slice to be lifted from the plate. When multiple bands are being excised from a DD gel, sterile technique should be adopted to avoid cross contamination between candidate cDNAs. For example, the location of the band on the gel should be marked with a sterile needle for each band and the appropriate gel slice removed using a fresh scalpel blade each time.

To facilitate reamplification (Section 5.2), the gel slice containing the candidate band should be removed directly into the tube to be used for reamplification. This avoids subsequent transfer between tubes and also reduces the risk of cross contamination. It is also a good precautionary measure to cut the original gel slice in half and store one half in a PCR tube at $-20\,^\circ$C in case of a failed reamplification reaction. Once the gel slice has been isolated, we have routinely stored them before reamplification for over three months at $-20\,^\circ$C. Upon recovery of a gel slice, the remaining differential display gel should be re-exposed to ensure that the correct band has been selected. If more than the candidate band has been removed, this will give an indication of the number of unique sequences to expect upon the analysis of subclones. If the homogeneity of the target band is to be checked using SSCP analysis (29) then the band adjacent to the target band must also be recovered from the gel.

Protocol 4

Candidate band excision directly from glass plate

Equipment and reagents

- Sterile needle and scalpel blade
- PCR tubes
- RNase-free sterile water

Method

1. Align the autoradiogram with the exposed image to the glass plate using the characteristic outline of the gel (or radioactive markers) and secure with adhesive tape.

2. Define the boundaries of the target band using a sterile needle. Score the outline of the band using a scalpel blade on the dried gel and rehydrate the gel slice using 2 μl of sterile water. Excise the rehydrated gel slice (~ 2 mm) using the same scalpel blade.

3. Place the gel slice into sterile PCR tube containing 2 μl sterile water and allow to rehydrate fully for 20 min (final volume ~ 5 μl). Spin the tube briefly to deposit the gel slice at the bottom of the tube.

4. Once the excised gel slice has been fully rehydrated, store at −20°C, or use directly in the reamplification protocol.

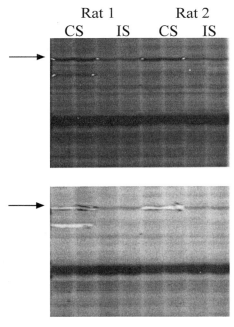

Figure 3 A putative band (indicated by the *arrow* in the top panel) was identified as potentially down-regulated in ipsilateral striatum (IS) compared to contralateral striatum (CS) after unilateral 6-OHDA-induced lesioning in two independent rats (rats 1 and 2). Following careful rehydration of the relevant gel slice, precise excision of the appropriate fragment was confirmed by re-exposing the dried gel to film (indicated by the *arrow* in the lower panel).

5.2 Reamplification of candidate bands

To obtain sufficient material for downstream analysis, the candidate cDNA must be reamplified from the excised gel slice (*Protocol 5*). We have reamplified excised fragments using both the original differential display anchor and arbitrary primer combinations, or a full-length T7 promoter primer (5′-GTAATACG-ACTCACTATAGGGC-3′) and a full-length M13 reverse (–48) primer (5′-AGCGGA-TAACAATTTCACACAGGA-3′). Both methods were successful in generating reamplification products of sufficient abundance for downstream cloning. Where the target band is to be reamplified with the original differential display primer pair, it is important to use the same MgCl$_2$ concentration as that used to generate the original differential display pattern. The advantage of using the full-length T7 and M13 primers is to facilitate riboprobe synthesis if required, but also to simplify the reamplification of numerous candidate bands simultaneously, since these primers are suitable for every candidate band, allowing the preparation of a core master mix. Using each individual primer set specific to each candidate for reamplification makes the PCR assay rather more complicated.

Protocol 5

Reamplification of candidate bands

Equipment and reagents

- PCR tubes or plates
- Thermal cycler (e.g. PTC-200, MJ Research)
- 2 µM 5′ primer: original arbitrary primer or full-length M13 (–40) primer
- 2 µM 3′ primer: original anchored primer or full-length T7 primer
- 25 mM MgCl$_2$ (Perkin Elmer)

- 10 × PCR buffer II: 100mM Tris–HCl pH 8.3, 500 mM KCl (Perkin Elmer)
- RNase-free sterile water
- 250 µM dNTP mix (Hieroglyph kit)
- 5 U/µl AmpliTaq polymerase (Perkin Elmer)
- Rehydrated gel slice

Method

1. Whilst the gel slice is rehydrating (see *Protocol 4*), or thawing out if previously frozen, prepare the reamplification mix as follows using the original differential display primers (per reaction). (N.B. If using full-length T7 and M13 (–40) primers a core master mix for the number of reactions (+10%) can be prepared.)
   - 17.8 µl sterile water
   - 4 µl of 10 × PCR buffer
   - 1.2 µl MgCl$_2$
   - 3.2 µl dNTP mix
   - 4 µl 5′ primer
   - 4 µl 3′ primer
   - 0.8 µl AmpliTaq polymerase

Protocol 5 continued

2. Add 35 μl reamplification mix to 5 μl rehydrated gel slice and carry out the PCR using the following cycling parameters:

   (a) 95 °C for 5 min.

   (b) 30 cycles of 94 °C for 15 sec, 60 °C for 30 sec, 72 °C for 2 min.

   (c) 72 °C for 7 min.

   (d) Hold at 7 °C.

3. Combine 15 μl of the PCR product with 2 μl of bromophenol blue DNA loading dye (10 ×) and run on a 1.5% agarose gel.

Using the *Protocol 5* described here, we were regularly able to reamplify candidate bands of 200–1600 bp (*Figure 4*). In all cases the specificity and abundance of the reamplified products was sufficiently high to allow rapid subcloning (Section 5.4). We have occasionally experienced reamplification problems, possibly due to incomplete coverage of the gel segment by PCR buffer. Alternatively, if the gel slice was too large, or the gel was not washed sufficiently, excess urea and acrylamide may inhibit the reaction, but this was overcome by increasing the reaction volume to 100 μl. Doubling the amount of AmpliTaq polymerase also helps, as well as using other polymerases such as AmpliTaq Gold (Perkin Elmer). If insufficient reamplification product is obtained then a 2 μl aliquot of

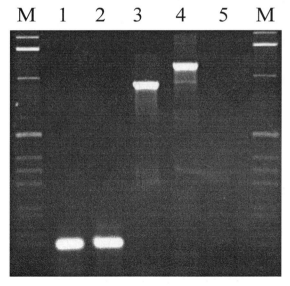

Figure 4 Reamplification of putative differentially expressed cDNAs. Using *Protocol 5*, both short (lanes 1 and 2) and long (lanes 3 and 4) cDNA fragments were successfully reamplified directly from the gel slice. In all cases the target PCR product was the most abundant. Lane 5 is the negative control where the band to be reamplified was replaced by sterile water. M is the 1 kb ladder (molecular weight marker).

the purified reamplification reaction can be reamplified again using the same PCR parameters as the first round of reamplification. However, as with all PCR reactions, it is imperative to include a negative control where the gel slice is replaced by sterile water. When the putative differentially displayed PCR product of interest is greater than 2 kb it may be necessary to increase the elongation times of the PCR and use long distance PCR enzyme systems.

5.3 Purification of reamplified products

Prior to the cloning, it is necessary to isolate the target PCR product from other components of the reamplification mixture, in particular any primer dimer artefacts. The type of purification required depends upon the specificity of the original reamplification reaction. Where there is a single reamplification product we routinely column purify the entire reaction using Wizard PCR DNA Purification System as per manufacturer's instructions (Promega). However, if other non-target PCR bands exist (primarily distinguished upon the basis of a different size to the target band), then it is necessary to gel purify the target band using the Geneclean Spin kit (Bio 101 Inc.) or other similar gel isolation methods (e.g. Qiaquick gel extraction kit, Qiagen). Co-migrating species increase in proportion during reamplification and may also clone more efficiently so this step is very important. Depending on the abundance, resuspend the PCR product in 10–20 μl.

5.4 TA cloning reamplified candidate bands

Previously the cloning of PCR products was notoriously difficult, requiring either the addition of restriction endonuclease sites into the primers or blunt-end treatment of the amplification product and ligation into a suitable vector. However, with the advent of 'TA cloning' the cloning of PCR products has become routine. TA cloning relies upon the terminal transferase activity of *Taq* polymerase which adds a single deoxyadenosine (A) to the 3′ ends of PCR products. Cloning of the PCR product can be achieved by ligating into a linearized vector which has a single overhanging 3′ deoxythymidine (T) residue. In our experience, the TA cloning procedure (TA cloning kit, Invitrogen) has proved extremely reliable for the cloning of reamplified differential display cDNAs independent of the method of purification of the PCR product.

Recently, to facilitate the rapid cloning and analysis of PCR products we have used a method based on the TOPO TA cloning system (Invitrogen, USA), which exploits the ligation activity of topoisomerase. The advantage of this system is that PCR products can be successfully ligated into a modified TA vector in 5–10 min with an efficiency comparable to standard T4 DNA ligase-mediated cloning. When cloning PCR products using the TA strategy it is always necessary to purify the target band from the rest of the reaction (as in Section 5.3) to ensure that any primer dimer artefacts are not preferentially ligated into the TA vector.

In our experience either newly reamplified and purified PCR products, or purified PCR products that have been previously frozen, can be successfully ligated into TA vectors with similar efficiency. Whilst the TOPO TA cloning system is extremely efficient, we have observed that the topoisomerase/TA vector used is very thermolabile, with the efficiency of ligation being substantially reduced each time the vector experiences a change in temperature. Therefore it is advisable to use up each aliquot of vector in one experiment (i.e. perform several ligations), to avoid reduced ligation efficiency. The efficiency of ligation into TA vectors is inversely proportional to the increasing size of the target PCR product although using *Protocol 6* we have routinely cloned differential display PCR products up to 2 kb. When using TA vectors it is advisable to include a 'no insert' control. Occasionally, when candidate bands fail to clone using the TA system, alternative blunt-end cloning strategies should be adopted.

Protocol 6

Cloning reamplified PCR products

Equipment and reagents

- TOPO TA cloning kit (Invitrogen) including TOP10 One Shot competent cells
- 40 μg/ml X-gal (Sigma)
- LB plates containing 50 μg/ml ampicillin (Sigma)
- Ampicillin (Sigma)

Method

1. Combine 4 μl of the resuspended PCR product (generated in Section 5.3) with 1 μl of topoisomerase/TA vector mixture, and incubate at room temperature for 15 min.
2. Whilst the PCR product is ligating, defrost the competent cells on ice and add 2 μl of 2-mercaptoethanol to each aliquot.
3. Add 3 μl of the ligation reaction to the cells and incubate on ice for 30 min.
4. Heat shock the transformation reaction at 42°C for 30 sec and return the tubes to ice for 5 min.
5. Add 200 μl of SOC media to the transformation reaction and 'recover' the cells by placing the tubes in a shaking incubator at 37°C for 30 min.
6. Plate out the transformation reaction onto LB agar plates containing 50 μg/ml ampicillin and 40 μg/ml X-gal. There is no need to add IPTG to the LB agar plates when using the TOP10 cells since they do not express the *lac*I repressor.
7. Leave the plates to dry for 10 min at room temperature prior to inverting them and incubating at 37°C for 16 h.
8. Place the plates at 4°C for about 2 h to develop the blue colour of colonies which do not contain an insert.

5.5 Colony selection and subclone identification

The purpose of colony selection is to identify the actual differentially expressed cDNA from the background of any other unique sequences which have been reamplified and cloned unintentionally. A reamplified differential display band may consist of several unique sequences and this is a prime reason for false positive generation. Undoubtedly, the most common cause of contamination is the excision of multiple bands, rather than just the target band from the dried gel. Even when all the precautions have been adopted, any cDNAs that co-migrate with the putative differentially expressed band will also be amplified. However, given that the target band has been selected on the basis of a visual change in gene expression, then kinetically it should represent the most abundant product in the final reaction. This situation is complicated somewhat by the fact that 'background' sequences may amplify or clone with a greater efficiency than the target band. As such, it is necessary to analyse several subclones of the reamplified product. Ideally, a minimum of eight independent colonies should be selected and their inserts sequenced or restriction mapped (after preparation of plasmids using miniprep kits such as Qiaspin from Qiagen), to give an indication of the number of unique sequences in the original reamplification. Alternatively, subclones can be selected by PCR screening direct from the LB plates, using primers flanking the cloning site. In general, for all projects described, the clone that was the most highly represented turned out to be the actual differentially expressed cDNA.

For the projects described here, the identity of at least eight inserts for a given target band were screened using ABI 377 automated sequencing technology (Perkin Elmer). However, an alternative is to screen inserts using random frequent-cutter restriction enzymes. If two or more independently cloned reamplified products share identical restriction patterns (given that the restriction enzyme of choice cuts internally), then statistically they are likely to represent the same sequence. Therefore, in these cases only one out of the group of matching clones needs to be sequenced, thereby reducing the amount of sequencing reactions required. This approach also gives an indication of the number of unique sequences in the original reamplified candidate band. *Figure 5* shows a restriction pattern of eight independent subclones from the same target band which have been digested with *Eco*RI. Seven out of eight subclones showed the same restriction pattern (one showed false positive insert) which suggests that the target sequence is unique in the reamplified PCR product. When sequenced, all seven clones were confirmed to be the same sequence.

In all differential display projects we have identified a high percentage of sequences that show no matches with any sequences in the public databases (e.g. GenBank). Since differential display is a technique utilizing cDNA synthesis primed by oligo(dT) anchored primers annealing to the poly(A) tail of mRNA, many candidate sequences are in the 3′ UTR often making their identity difficult to confirm. However, since we have been isolating larger cDNA fragments using the Hieroglyph primers in combination with the GenomyxLR electrophoresis system, we tend to identify more coding information than we did previously. For these totally unknown genes we have pursued the 5′ rapid amplification of cDNA ends assay (5′ RACE, Clontech) to obtain coding sequence and therefore increase the possibilities of hitting a database match. Doublet bands were frequently observed using denaturing polyacrylamide gels and were

M 1 2 3 4 5 6 7 8 9 10 11 12 M

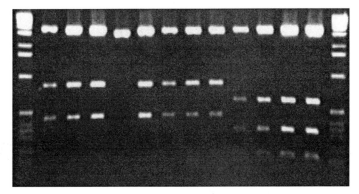

Figure 5 Restriction endonuclease pattern of eight (lanes 1–8) independent subclones from the same target band which have been digested with *Eco*RI. Seven of the eight subclones showed the same restriction pattern suggesting the target sequence was unique in the reamplified PCR product. Lane 4 was a false positive subclone containing no insert. Additionally, similar analysis of four independent subclones (lanes 9–12) from a different target band revealed four inserts with the same digest pattern. Sequence analysis of all the subclones confirmed that the identity of the subclones for each target band were identical.

usually identified as the same species when sequenced. We also confirmed the presence of primer redundancy, since we sequenced the same gene more than once, despite the original candidate band being identified using different primer combinations.

6 Confirmation of differential expression using independent techniques

When a large difference is observed between test samples on a differential display gel the subsequent confirmation using more quantitative techniques often shows that the difference is much less dramatic than expected, demonstrating the essentially qualitative nature of the differential display technique (30). For example, in our HT4 neuronal cell culture model, we identified two putative differentially expressed cDNAs of similar intensity and banding pattern. However, on confirmation using TaqMan™ real time quantitative PCR, one was confirmed as four- to sixfold difference, whilst the other was a marginal difference, despite these differences looking similar on the original gels.

Putative differential display hits generated in our laboratories have been confirmed using a variety of techniques depending on the experimental model used and the practicalities of the confirmation method. For the cell culture model, since enough RNA was available, we started with Northern blot analysis of putative hits generated from the differential display screen. We found this rather time-consuming and a little unreliable, especially when using certain digoxygenin (DIG) labelling techniques. We attempted DIG incorporation during

PCR with plasmids containing the differentially expressed cDNA of interest, but this was quite inconsistent. Utilizing DIG end-labelling of 30-mer oligonucleotides (50% GC content) was much more successful, although some mRNAs identified by differential display were beyond the level of detection by Northern blotting. For these we utilized the TaqMan™ quantitative PCR technique. *Figure 6* shows a candidate band identified as down-regulated in RC and RCG lanes, which was subsequently confirmed using both Northern blotting and TaqMan™ analysis.

Protocol 7

TaqMan™ real time quantitative PCR confirmation

Equipment and reagents

- cDNA samples (prepared as in *Protocol 8*)
- Plasmid or genomic DNA samples for standard curve generation
- 10 × TaqMan™ buffer A (TaqMan™ core reagents kit, Perkin Elmer)
- 25 mM $MgCl_2$ (Perkin Elmer)
- 10 mM dATP, dGTP, dCTP and 20 mM dUTP (Perkin Elmer)
- 10 μM sense gene-specific primer

- 10 μM antisense gene-specific primer
- 5 μM TaqMan™ gene-specific probe (FAM labelled, Perkin Elmer)
- 1 U/μl AmpErase uracil N-glycosylase (Perkin Elmer)
- 5 U/μl AmpliTaq Gold (Perkin Elmer)
- ABI Prism 7700 sequence detector (Perkin Elmer)

N.B. TaqMan™ probe and primers designed using Primer Express software (Perkin Elmer) and synthesized by Perkin Elmer.

Method

1. For each target gene, prepare core master mix (+10% volume) of the following reagents (per reaction, including standard curve):
   - 11.625 μl sterile water
   - 2.5 μl TaqMan™ buffer
   - 6 μl $MgCl_2$
   - 0.5 μl each of dATP, dCTP, dGTP, and dUTP
   - 0.25 μl UNG
   - 0.5 μl TaqMan™ probe
   - 0.5 μl sense primer
   - 0.5 μl antisense primer
   - 0.125 μl AmpliTaq Gold
2. Add 24 μl master mix to each well of a TaqMan™ plate.
3. Add 1 μl cDNA prepared from a reverse transcription reaction of 1 μg RNA in 20 μl to the appropriate wells.

4. Add 1 μl of plasmid DNA (400 pg) or genomic DNA(4 pg) to the appropriate standard curve wells for the top dilution and 1 μl of 100-fold dilutions into other appropriate wells.

5. Place the plate into the ABI Prism 7700 and run the PCR with the following parameters:

    (a) 50 °C for 2 min.

    (b) 95 °C for 10 min.

    (c) 40 cycles of 95 °C for 15 sec and 60 °C for 1 min.

6. Analyse data using Power Macintosh software supplied with the ABI Prism 7700.

Whole embryo *in situ* hybridization studies were most preferable for the chick embryo project due to very small amounts of starting material. *In situ* hybridization was carried out on whole mount chick embryos with gene-specific riboprobes using standard techniques (31). *Figure 7* shows confirmation of a candidate band present in R1 and MB but not in HB.

We utilized standard RT-PCR as a first stage confirmation of differential expression in rat neuropathic pain model with the view to subsequently screening putative hits in TaqMan™ quantitative PCR plates containing multiple tissues and time points. Following cloning and sequencing, the most abundant clone species was identified and specific primers designed internal to the original set used for differential display. The usual criteria for primer selection were followed, utilizing commercial primer design software to give primers that were generally 18–21 nucleotides long with 50–60% GC content. It is important to ensure that the primer regions selected do not have any ambiguous base designations, therefore the sequences selected should be the same in each clone for a given clone species. Relative RT-PCR was then carried out using equal amounts of total RNA from the samples to be compared, under identical reaction conditions. The products were then analysed by agarose gel electrophoresis with the relative amounts of product from each reaction proportional to the relative abundance of the RNA transcript in the original RNA samples. Depending on the abundance of the targeted cDNA, cycle and cDNA dilution studies may need to be performed to avoid saturating PCR conditions. *Figure 8* shows the results of a typical RT-PCR confirmation experiment using specific primers for the most abundant clone of a cDNA fragment as well as primers for a less abundant species. It is evident that the most abundant clone was differentially expressed. In the rat neuropathy study, the clones that represented the majority species for a given candidate cDNA fragment were subsequently demonstrated to be differentially expressed in nine out of ten instances.

Finally for the 6-OHDA-lesioned rat study, hits that were generated in two individual rats were confirmed using TaqMan™ real time quantitative PCR (*Protocol 7*) in another two rats independent from the original differential display

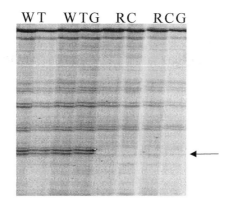

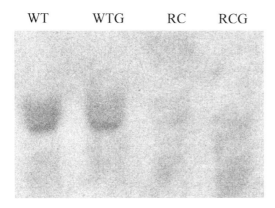

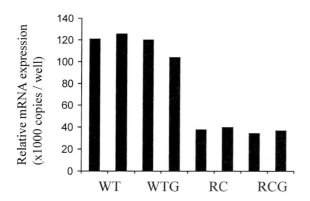

Figure 6 A putative differentially expressed cDNA was identified as down-regulated in RC/RCG lanes in the HT4 neuronal cell line model (upper panel). This differential expression was subsequently confirmed using Northern blotting with a gene-specific DIG end-labelled 30-mer oligonucleotide (middle panel), and also TaqMan™ real time quantitative PCR using gene-specific primers and TaqMan™ probe (lower panel). We gratefully acknowledge Laura Smith and Amanda Barton for the Northern blot.

Protocol 8

RT-PCR confirmation

Equipment and reagents

- Thermal cycler (e.g. PTC-200, MJ Research)
- Sterile PCR tubes or plates
- SuperScript II RNase H⁻ reverse transcriptase (200 U/μl) supplied with 5 × first strand SuperScript II buffer (250 mM Tris–HCl pH 8.3, 375 mM KCl, 15 mM $MgCl_2$) and 100 mM DTT (Gibco BRL)
- 10 mM dATP, dGTP, dCTP, and dTTP (Perkin Elmer)
- 0.1 μg/μl oligo(dT)$_{12-18}$ primer (Gibco BRL)

- 40 U/μl RNaseOUT (Gibco BRL)
- RNase-free sterile water
- Total RNA template (diluted to 250 ng/μl)
- 10 × PCR buffer
- 25 mM $MgCl_2$
- 2 μM gene-specific antisense primer
- 2 μM gene-specific sense primer
- 5 U/μl AmpliTaq polymerase

Method

1. Thaw RNA samples on ice. If possible use separate sources of RNA to that employed for the original differential display to control for differences unrelated to the experimental condition under investigation.

2. Add 2 μl oligo(dT)$_{12-18}$ primer to 2 μl total RNA (0.5 μg) and incubate at 65 °C for 5 min using a thermal cycler with a heated lid, then quickly transfer to ice.

3. Prepare reverse transcription mixes on ice in sufficient volume for the number of reactions required (+10%). Also prepare a separate core master mix with water for minus RT controls. All volumes given are per reaction.
   - 4 μl Superscript II buffer
   - 0.5 μl each of dCTP, dATP, dTTP, and dGTP
   - 2 μl DTT
   - 0.5 μl RNaseOUT
   - 1 μl Superscript II
   - 6.5 μl RNase-free sterile water

4. Add 16 μl of master mix per tube and carefully mix by pipetting.

5. Perform first strand cDNA synthesis using a thermal cycler with a heated lid, at 42 °C for 50 min, followed by inactivation of the reaction at 70 °C for 15 min.

6. Add 30 μl sterile dH$_2$O per tube and mix.

7. Add 2 μl of cDNA templates prepared above to PCR tubes. Also include a tube with 2 μl of sterile water as a minus template control.

8. Prepare a PCR master mix as follows (volume per reaction):
   - 14.3 μl RNase-free sterile water
   - 3 μl 10 × PCR buffer
   - 2.4 μl $MgCl_2$

Protocol 8 continued

- 0.5 µl each of dATP, dGTP, dTTP, and dCTP
- 3 µl antisense primer
- 3 µl sense primer
- 0.3 µl AmpliTaq polymerase

9. Add 28 µl of master mixture per tube and mix by pipetting.

10. Perform PCR using the following parameters:

    (a) 95°C for 2 min.

    (b) 30 cycles of 92°C for 30 sec, 58°C for 30 sec, 72°C for 1 min. N.B. Cycle studies should be performed to define the linear range of amplification and therefore the appropriate cycle number. The annealing temperature may need to be altered depending on the T_m of the primers.

11. Add 6 µl of 6 × gel loading buffer to each tube and analyse 10 µl by agarose gel electrophoresis.

mb   r1   r2-7

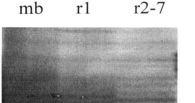

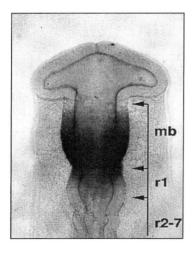

Figure 7 Whole mount *in situ* hybridization on a stage 10–12 chick embryo of a putative differentially expressed cDNA. Analysis of the staining pattern revealed that the candidate band was expressed in the midbrain (mb) and rhombomere 1 (r1) but not in hindbrain (r2–7). This was the pattern expected from the initial differential display gel. We gratefully acknowledge Huma Sheikh for this *in situ* image.

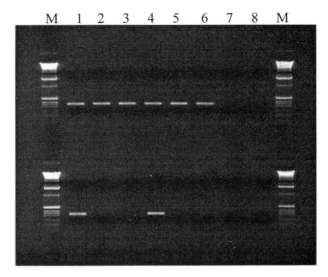

Figure 8 RT-PCR analysis using specific primers for the most abundant clone of a differentially expressed band (bottom row) and primers for a less abundant clone of the same band (top row). The differentially expressed clone therefore corresponds to the most abundant clone. RT-PCR was performed using mRNA from ipsilateral (lanes 1 and 4) and contralateral (lanes 2 and 5) rat DRG following unilateral partial ligation. Lanes 3 and 6 are RT-PCR reactions using DRG from sham-surgery rats. Lanes 7 and 8 are no reverse transcription controls using mRNA from ipsilateral (lane 7) and contralateral (lane 8) DRG of nerve-injured animals.

analysis. *Figure 9* shows successful confirmation of a differentially expressed cDNA up-regulated in striatum ipsilateral to the lesion in both animals.

References

1. Liang, P. and Pardee, A. B. (1992). *Science*, **257**, 967.
2. Callard, D., Lescure, B., and Mazzolini, L. (1994). *BioTechniques*, **16**, 1096.
3. Ayala, M., Balint, R. F., Fernandez-de-Cossio, M. E., Canaan-Hadan, L., Larrick, J. W., and Gavilondo, J. V. (1995). *BioTechniques*, **18**, 842.
4. Liang, P. and Pardee, A. B. (1995). *Curr.Opin. Immunol.*, **7**, 274.
5. Douglas, S. A., Averboukh, L., and Ohlstein, E. H. (1996). *Pharmacol. Rev.Commun.*, **8**, 267.
6. Wang, X., Yue, T., Barone, F. C., Ruffalo, Jr. R. R., and Feuersein, G. Z. (1994). In *Pharmacology of cerebral ischemia* (ed. J. Krieglstein and H. Oberpichler-Schwenk), p. 473. Wissenschaftliche Verlagsgesellschaft mbH, Stuttgart.
7. Liang, P., Averboukh, L., Keyomarsi, K., Sager, R., and Pardee, A. B. (1992). *Cancer Res.*, **52**, 6966.
8. Vogeli-Lange, R., Burckert, N., Boller, T., and Wiemken, A. (1996). *Nucleic Acids Res.*, **24**, 1385.
9. Morimoto, B. H. and Koshland, D. E. (1990). *Neuron*, **5**, 875.
10. Choi, D. W. (1988). *Neuron*, **1**, 623.
11. Maher, P. and Davis, J. B. (1996). *Neuroscience*, **6**, 6394.
12. Thompson, C. B. (1995). *Science*, **267**, 1456.

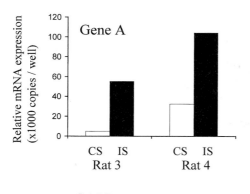

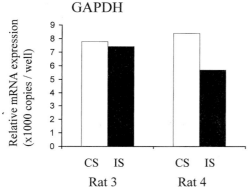

Figure 9 Confirmation using TaqMan™ quantitative PCR of a gene up-regulated in ipsilateral striatum (IS) compared to contralateral striatum (CS) after 6-OHDA lesioning in two different rats (rats 3 and 4) to those used for differential display analysis (upper panel). The lower panel shows results for GAPDH expression from the same cDNA pool showing little change in expression after lesioning.

13. Bronner-Fraser, M. and Fraser, S. E. (1997). *Curr. Opin. Cell. Biol.*, **9**, 885.
14. Gavalas, A., Studer, M., Lumsden, A., Rijli, F. M., Krumlauf, R., and Chambon, P. (1998). *Development*, **125**, 1123.
15. Seltzer, Z., Dubner, R., and Shir, Y. (1990). *Pain*, **43**, 205.
16. Na, H. S., Yoon, Y. W., and Chung, J. M. (1996). *Pain*, **67**, 173.
17. Nielsch, U. and Keen, P. (1989). *Brain Res.*, **481**, 225.
18. Mohiuddin, L. and Tomlinson, D. (1997). *Diabetes*, **46**, 2057.
19. Ungerstedt, U. (1968). *Eur. J. Pharmacol.*, **5**, 107.
20. Gerfen, C. R., Engber, T. M., Mahan, L. C., Susel, A., Chase, T. N., Monsma, F. J., *et al.* (1990). *Science*, **250**, 1429.
21. Zeng, B.-Y., Jolkkonen, J., Jenner, P., and Narsden, C. D. (1995). *Neuroscience*, **66**, 19.
22. Pollack, A. E. and Wooten, G. F. (1992). *Mol. Brain Res.*, **12**, 111.
23. Heid, C. A., Stevens, J., Livak, K. J., and Williams, P. M. (1996). *Genome Res.*, **6**, 986.
24. Dieffenbach, C. W., Lowe, T. M. J., and Dveksler, G. S. (1993). *PCR Methods Appl.*, **3**, S30.
25. Sheets, M. D., Ogg, S. C., and Wickens, M. P. (1990). *Nucleic Acids Res.*, **18**, 5799.
26. Liang, P., Averboukh, L., and Pardee, A. B. (1993). *Nucleic Acids Res.*, **21**, 3269.

27   Joshi, C. P. and Nguyen, H. T. (1996). *Plant Mol. Biol.*, **31**, 575.

28. Jurecic, R., Nguyen, T., and Belmont, J. W. (1996). *Trends Genet.*, **12**, 502.

29. Mathieu-Daude, F., Cheng, R., Welsh, J., and McClelland, M. (1996). *Nucleic Acids Res.*, **24**, 1504.

30. Bauer, D., Muller, H., Reich, J., Riedel, H., Ahrenkiel, V., Warthoe, P., *et al.* (1993). *Nucleic Acids Res.*, **21**, 4272.

31. Shamim, H., Mahmood, R., and Mason, I. (1998). In *Methods in molecular biology: molecular embryology* (ed. P. T. Sharpe and I. J. Mason) p. 623. Humana Press, NJ.

Chapter 4

RAP-array: expression profiling using arbitrarily primed probes for cDNA arrays

BARBARA JUNG, THOMAS TRENKLE, MICHAEL McCLELLAND, FRANCOISE MATHIEU-DAUDE, and JOHN WELSH

Sidney Kimmel Cancer Centre, 10835 Altman Road, San Diego, CA 92121, U.S.A.

1 Introduction

We present a method for detecting differential gene expression using RNA arbitrarily primed PCR (RAP-PCR) fingerprints as probes against cDNA arrays. Using inexpensive arrays of cDNA clones, together with reduced complexity probes derived from RAP-PCR, this approach makes cDNA array technology generally accessible to any laboratory. This hybrid method combining RNA fingerprinting with differential hybridization is many times faster than the standard gel analysis of RNA fingerprints. We present an example where a RAP-PCR fingerprint (*Figure 1*) was converted to probe and used to address an array of approximately 18 000 IMAGE consortium cDNA clones by differential hybridization (*Figures 2* and *3*). In this chapter, we describe protocols for preparing probes from RNA fingerprint reactions and for using them in differential screening against large cDNA arrays.

Inexpensive arrays of the IMAGE consortium ESTs such as those shown in *Figure 2* and *3* containing 18 432 clones in *E. coli*, double-spotted on 22 × 22 cm membranes, are commercially available, at a cost of about $200 per membrane (Genome Systems). Membranes with hundreds or thousands of PCR products from cDNA clones are also commercially available (Clontech and Research Genetics).

In general, detecting clones from the most abundant few thousand mRNAs in a cell using a radiolabelled probe containing the full complexity of the mRNA population is not difficult (1), but background hybridization limits the detection of rarer mRNAs. To address these limitations, different support materials, direct synthesis of oligos in an array using photolithography, and robotic deposition in very dense arrays, have been explored. PCR products from segments of mRNAs which are attached to glass at high density can be custom-made for any library but are currently not readily available (2, 3). When using fluorescent dyes, each cDNA or mRNA population to be used as a probe can be labelled with a different fluorescent dye. Depending on the fluoroimager, one to five different dyes can be assessed on one array after hybridization (4–6). Alternatively, oligonucleotides have been attached to either a glass or silicon surface, or manufactured by sequential photochemistry on a chip (7) with tens of thousands of different oligo sequences per square centimetre. Arrays of oligo nucleic acid analogues such as peptide nucleic acids are also under development (8).

Detection sensitivity of as few as one copy of a mRNA per cell with a dynamic range of 10 000-fold has been reported (2). In practice, however, achieving this sensitivity remains a challenge. In addition to dynamic range and sensitivity, other important issues for many researchers who want to use array technology are cost and accessibility.

In this chapter, we demonstrate a strategy to improve the signal of rarer messages in cDNA arrays by using RAP-PCR for probe preparation. A RAP-PCR probe consists of sequences represented in a total mRNA population, including rarer transcripts, but with altered abundance ratios relative to a total mRNA-derived probe. Generating probes using RAP-PCR improves the sensitivity and dynamic range for subsets of cDNAs. By altering the abundance ratios of sequences in the probe, a subset of those mRNAs that are so rare that they are difficult to detect with probes generated from total message are sufficiently represented to be easily detected on cDNA arrays of colonies. It was shown previously that decreasing the complexity of a differential hybridization probe by subtraction allows detection of transcripts at an abundance as low as 0.0005%, which is comparable to the sensitivity of confocal scanned chips (9). RAP-PCR probes are similar to subtracted probes in that some sequences are highly represented that are not normally highly represented in total mRNA. RAP-PCR probes differ from subtracted probes in that the highly represented sequences are arbitrary. The RAP-array method improves the signal-to-noise ratio for rarer transcripts, and does not attempt to examine all mRNAs simultaneously. This may also improve sampling of rare messages. Depending on the primers used, different subsets of the mRNA population are emphasised and greater coverage of the mRNA population is achieved by iteration of the method, as in RAP-PCR or differential display. However, the throughput of RAP-array is much higher than that of either RAP-PCR or differential display, making complete or nearly complete coverage realistic. In addition, RAP-PCR probes may have reduced complexity in that sequences that are represented in the RNA are not necessarily represented in the probe. We have referred to these probes as 'non-stoichiometric reduced complexity probes in the past (18).

We performed RAP-PCR on murine thymocytes that were either untreated or treated with the chemopreventive drug sulindac (see *Figure 1*) to detect differential gene expression. With the array approach, throughput is much greater than with the standard gel electrophoresis approach in that we are able to monitor at least 20 times as many transcripts per fingerprint reaction. Using the protocols described in this chapter, probes were generated from the finger-prints in *Figure 1* and hybridized to colony arrays (see *Figure 2* and *Figure 3*). We were able to detect about a thousand transcripts not visualized on the gel. In addition to accelerated throughput due to the higher resolution of the array detection method, 80% of the IMAGE clones have been partially sequenced, leading to a huge increase in throughput because cloning and sequencing are almost entirely avoided.

Finally, we will discuss the use of primers that have been chosen using the computer algorithm, *GeneUP*, to sample specific subsets of mRNA (10).

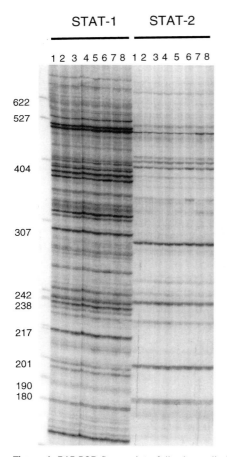

Figure 1 RAP-PCR fingerprints following sulindac treatment resolved on a polyacrylamide gel. Reverse transcription was performed with the arbitrary primer US6 (GTGGTGACAG) on 500 ng and 250 ng RNA. RNA was from untreated (lane 1, 2), sulindac (lane 3, 4), sulindac and cycloheximide (lane 5, 6), and cycloheximide-treated mouse thymocytes (lane 7, 8). RAP-PCR was performed with two primer pairs (STAT-1 forward/backward and STAT-2 forward/backward). Marker sizes in bp are indicated next to the left-hand lane. Note that no differences are detectable between the lanes with different template concentrations, or between the different treatments.

2 Generation of a RAP-PCR based reduced complexity probe to study differential gene expression on cDNA arrays

2.1 RNA fingerprinting

RNA arbitrarily primed PCR (RAP-PCR) using arbitrary primers (11, 12) has been used to identify and characterize differentially expressed genes under various conditions such as TGF-β treatment of multiple epithelial cell lines (13, 14), UV treatment of melanocytes (15, 16), or bile acid treatment of colon carcinoma cells (17). Now, we make further use of this method to generate a reduced complexity probe from a RNA fingerprint, which is labelled and hybridized to an array (18). The hybridization enables us to visualize products that cannot be evaluated with standard gel analysis of RAP-PCR products.

In RAP-PCR, first strand cDNA is generated with reverse transcriptase using an oligo(dT), a primer of arbitrary sequence, or a 3' anchored oligo(dT) primer. Next, in RAP-PCR, second strand cDNA is primed with another arbitrary primer and PCR takes place between the arbitrary primers (19). In differential display, PCR takes place between an arbitrary primer and the 3' anchor (20). In the example we presented in *Figures 1*, *2*, and *3* we used the statistical primers STAT-1 forward (AACCTCCCCAG) and STAT-1 backward (GGGGTCTTCAG) to target murine genes involved in major apoptotic pathways. Protocols are presented below (*Protocol 1* and 2).

It is very important to control for variation of cDNA products between RNA samples by performing RAP-PCR at two RNA concentrations differing by a factor of two or more (*Figure 1*). A small subset among the many cDNA products are very sensitive to amplification conditions and RNA concentration and quality. The use of two RNA concentrations allows these products to be identified and ignored in subsequent analyses (11, 12). The fingerprint is only suitable for hybridization to an array if concentration-dependent differences are minimal.

Protocol 1

RNA preparation

Equipment and reagents

- Table-top microcentrifuge (10 000 g)
- Spectrophotometer (260 nm and 280 nm)
- QIAshredder (Qiagen)
- RNase-free DNase and RNase inhibitor (Boehringer Mannheim)

- RNeasy Total RNA Purification Kit (Qiagen)
- 10 × TBE pH 8.3: 108 g Tris base, 55 g boric acid, 7.44 g EDTA in 1 litre dH$_2$O
- Agarose (e.g. Amnesco)

Method

1. Prepare RNA after harvesting up to 1×10^7 treated and untreated cells in 600 μl to 1.2 ml of lysis buffer (from RNeasy Kit) each.

2. Homogenize lysate through QIAshredder columns.

3. Prepare RNA using RNeasy Total RNA Purification Kit, eluting in 90 μl of RNase-free water.

4. Incubate RNA eluate in a total volume of 100 μl of 20 mM Tris, 10 mM MgCl$_2$ buffer with 0.08 U/μl of RNase-free DNase and 0.32 U/μl of RNase inhibitor for 40 min at 37°C. DNase treatment is important because small amounts of genomic DNA can contribute to the fingerprints.

5. Clean sample from enzymatic reaction using the RNeasy Kit eluting in 50 μl of RNase-free water.

6. Measure RNA quantity by spectrophotometry, and adjust RNA samples to 400 ng/μl in water. The expected yield varies from cell type to cell type, ranging from 25 μg to 120 μg.

7. Check for quality and concentration of RNA by agarose gel electrophoresis (1%) of 400 ng of RNA in 1 × TBE buffer. Store RNA samples at −80°C.

Protocol 2

RNA fingerprinting

Equipment and reagents

- Thermocycler (Perkin Elmer Cetus)
- Sequencing size gel electrophoresis apparatus
- BioMax film (Eastman Kodak)
- Arbitrary or statistical primer pair (10–20-mer): 100 μM stocks
- Oligo(dT) primer (15-mer) (Genosys Biotechnologies): 100 μM stock
- 2 × PCR mixture: 20 mM Tris pH 8.3, 20 mM KCl, 6.25 mM MgCl$_2$, 0.35 mM of each dNTP, 2 μM of each arbitrary primer, 0.1 μCi/μl [α-^{32}P]dCTP (ICN), and 0.25 U/μl AmpliTaq® DNA polymerase Stoffel fragment (Perkin Elmer Cetus)

- 2 × RT mix: 100 mM Tris pH 8.3, 150 mM KCl, 6 mM MgCl$_2$, 40 mM DTT, 0.4 mM of each dNTP, 1 μM of primer (either oligo(dT) or arbitrary primer), and 4 U/μl of M-MLV reverse transcriptase (Promega)
- Formamide/EDTA/XC/BPB gel loading buffer: 10 ml formamide (Aldrich), 10 mg xylene cyanol FF, 10 mg bromophenol blue, 200 μl 0.5 M EDTA pH 8.3
- 40% acrylamide stock solution (FisherBiotech)
- Urea
- 10 × TBE (see *Protocol 1*)

Method

1. Perform RT-PCR on total RNA using two concentrations per sample (500 ng and 250 ng per reaction, corresponding to 50 ng/μl and 25 ng/μl of RNA per RT reaction) and an oligo(dT) primer (15-mer) or an arbitrary primer (10–11-mer).

2. Combine RNA (5 μl) with 5 μl of 2 × RT mixture for a 10 μl final reaction.

Protocol 2 continued

3. Include a reverse transcriptase-free control in initial RAP-PCR experiments to check for DNA contaminants in RNA.

4. Perform reverse transcription at 37°C for 1 h (after a 5 min ramp from 25°C to 37°C), inactivate the enzyme by heating the samples at 94°C for 5 min, and dilute the newly synthesized cDNA fourfold in water.

5. Mix diluted cDNAs (10 μl) with the same volume of 2 × PCR mixture for a 20 μl final reaction, including a pair of two different oligonucleotide primers of arbitrary sequence. *In general, there are no particular constraints on the primers except that they contain at least a few C or G bases, that the 3' ends are not complementary with themselves or the other primer in the reaction to avoid primer dimers, and that primer sets are chosen that are different in sequence so that the mRNA population is resampled.*

6. Perform thermocycling using 35 cycles of 94°C for 1 min, 35°C for 1 min, and 72°C for 2 min.

7. Mix an aliquot of the amplification products (3.5 μl) with 9 μl of formamide dye solution, denature at 85°C for 4 min, and chill on ice. Load 2.4 μl onto a 5% polyacrylamide, 43% urea gel, prepared with 1 × TBE buffer. The PCR products resulting from the two different concentrations of the same RNA template are loaded side-by-side on the gel (see *Figure 1*).

8. Perform electrophoresis at 1700 V or at a constant power of 50–70 W until the xylene cyanol tracking dye reaches the bottom of the gel (approx. 4 h). Dry the gel under vacuum and expose to a Kodak BioMax X-ray film for 16–48 h.

Protocol 2 presents RAP-PCR using an oligo(dT) primer in the reverse transcription step and a pair of arbitrary primers in the PCR step. This is a modification of the protocol where an arbitrary primer was used in the reverse transcription step, followed by PCR with the arbitrary primer pair. Both methods work well. Alternatively, differential display can be performed using an oligo(dT) anchored primer during reverse transcription and then the anchored primer together with an arbitrary primer in the PCR (20). We are currently exploring the use of statistically selected primers chosen on the basis of their chance of occurrence in a target list. This strategy differs from the original RAP-PCR in that the primers are not selected arbitrarily. A program called *GeneUP* has been devised which uses a 'greedy' algorithm, which is an algorithm that searches information space starting in places where the information of interest is most likely to reside. In this program, the algorithm utilizes the likelihood of a match for each single primer to select primer pairs to sample sequences in the users list of interest (e.g. a list of human mRNAs associated with apoptosis). Sequences in another list (e.g. a list of abundantly expressed mRNAs in human cells and structural RNAs such as rRNAs, Alu repeats, and mtDNA) are simultaneously excluded (10). In principle, reduced complexity probes in which the messages of interest are more likely to be represented can be generated. *GeneUP* is freely available to non-commercial users. Another approach to sample specific genes of interest is to perform a PCR with a mixture of pairs of specific primers for each target gene under stringent conditions (multiplex-PCR) (21).

2.2 Probing of RNA fingerprints on cDNA arrays

The fingerprints generated using *Protocol 2* can then be used to generate probes for differential screening on arrays. First, however, it is wise to examine several major indicators of fingerprint quality. The fingerprints should be nearly identical at different starting RNA concentrations, which is ensured by comparing the fingerprints of different starting concentrations loaded side-by-side on a polyacrylamide gel (see *Figure 1*). There should be few significant concentration-dependent differences in the fingerprints, which could lead to serious misinterpretation of the data generated with the array. Only reproducible, concentration-independent fingerprints are chosen as probes for cDNA arrays. As RAP-PCR does not generate probes with a specific activity high enough to be used directly on an array, random primed labelling of the probe using *Protocol 3* is applied.

Protocol 3

Labelling of the cDNA probe

Equipment and reagents

- QIAquick Nucleotide Removal and PCR Purification Kit (Qiagen)
- 10 mM Tris–HCl pH 8.3
- 200 μM random hexamer oligonucleotide (Genosys Biotechnologies)
- 0.5 mM mix of dATP, dTTP, and dGTP
- 2.5 mM dCTP

- 3000 Ci/mmol [α-^{32}P]dCTP (ICN Pharmaceuticals)
- 10 × Klenow fragment buffer: 500 mM Tris–HCl pH 8.0, 100 mM MgCl$_2$, 500 mM NaCl
- Klenow fragment: 3.82 U/μl (Gibco BRL Life Technologies)

Method

1. Purify up to 2 μg of PCR product from RAP-PCR corresponding to approximately half of the PCR reaction using the QIAquick PCR Purification Kit in order to remove unincorporated bases, primers, and primer dimers under 40 base pairs. Recover the DNA in 50 μl of 10 mM Tris.

2. Combine 10% of the recovered fingerprint DNA (typically about 200 ng in 10 μl) with 3 μg random hexamer oligonucleotide primer, and 0.3 μg of each of the arbitrary RNA fingerprint primers in a total volume of 14 μl, boil for 3 min, and then place on ice.

3. Add the hexamer/primer/DNA mix to 11 μl reaction mix to yield a 25 μl reaction containing 0.05 mM of three dNTPs (minus dCTP), 100 μCi of 3000 Ci/mmol [α-^{32}P]dCTP (10 μl), 1 × Klenow fragment buffer, and 8 U Klenow fragment. Perform labelling reaction at room temperature for 4 h. For maximum probe length, chase the reaction by adding 1 μl of 2.5 mM dCTP and incubate for 15 min at room temperature, followed by an additional incubation of 15 min at 37 °C.

4. Remove the unincorporated nucleotides and hexamers with the Qiagen Nucleotide Removal Kit and elute the purified products in 280 μl, eluting twice in 140 μl 10 mM Tris pH 8.3 each time.

2.3 cDNA from poly(A)$^+$ mRNA as probe on cDNA array

To monitor the level of expression of the most abundant mRNAs in a total RNA preparation simultaneously, a probe from poly(A)$^+$ mRNA may be generated. The disadvantage of this technique is that rare messages are not be detectable on colony arrays because of overlapping background hybridization. It has been reported that at best 20% of the genes on an array of PCR products are sampled using a probe generated from total mRNA (1). This percentage can be expected to be significantly less on a colony array due to background hybridization. We have previously shown that 95% of the clones detectable after hybridization with a RAP-PCR fingerprint probe differ from those obtained after hybridization with a probe generated from poly(A)$^+$ mRNA. However, there are a number of circumstances where one is only interested in the genes that are most abundantly expressed, in which case a cDNA probe generated from poly(A)$^+$ RNA is suitable.

Protocol 4

Labelling a cDNA probe from poly(A)$^+$ mRNA

Equipment and reagents

- Table-top microcentrifuge (10 000 g)
- Water-baths at 70 °C, 50 °C, and 37 °C
- Oligotex mRNA Mini Kit (Qiagen)
- Random hexamer primers: 400 μM stock solution (Genosys Biotechnologies)
- 5 × AMV reaction buffer: 250 mM Tris–HCl pH 8.5, 40 mM MgCl$_2$, 150 mM KCl, 5 mM DTT
- 20 × dNTP mix: 33 mM each dGTP, dTTP, dATP

- AMV reverse transcriptase (Boehringer Mannheim)
- 3000 Ci/mmol [α-^{32}P]dCTP
- Master mix: 10 μl of 5 × AMV reaction buffer, 1 μl of 20 × dNTP, 2 μl AMV reverse transcriptase, and 10 μl of [α-^{32}P]dCTP in a final volume of 50 μl
- 33 mM dCTP
- Qiagen Nucleotide Removal Kit (Qiagen)
- 10 mM Tris pH 8.3

Method

1. Prepare poly(A)$^+$ RNA using Oligotex mRNA Mini Kit from total RNA prepared using *Protocol 1*.

2. Incubate 1 μg of poly(A)$^+$ mRNA and 9 μg of random hexamer primers in a volume of 27 μl at 70 °C for 2 min, and chill on ice.

3. Mix the RNA/hexamer mix with 23 μl of the master mix. Incubate reaction at room temperature for 15 min, ramp for 1 h to 47 °C, and hold at 47 °C for 1 h.

4. Chase the reaction with 1 μl of 33 mM dCTP for another 30 min at 47 °C.

5. Remove the unincorporated nucleotides and hexamers with the Qiagen Nucleotide Removal Kit and use two elution steps of 140 μl each, eluting the purified products in a final volume of 280 μl of 10 mM Tris pH 8.3.

Alternatively, we have applied a modified protocol using 'coding sequence' (CDS) primers (Clontech). In this case, the RNA/primer solution (1 μg of mRNA in 1 μl and 1 μl of 10 × CDS primers) is pre-heated to 70°C for 2 min. Then, the reaction is mixed with 8 μl master mix containing 2 μl of 5 × M-MLV reaction buffer (250 mM Tris–HCl pH 8.5, 40 mM MgCl$_2$, 150 mM KCl), 1 μl of 10 × dNTP mix, 1 μl M-MLV reverse transcriptase (Boehringer Mannheim), 0.5 μl of 100 mM DTT, and 3.5 μl of [α-^{32}P]dCTP in a final volume of 10 μl. The reaction is incubated at 50°C for 20 min, is chased following *Protocol* 4, step 4, and purified following *Protocol* 4, step 5.

3 Arrays for studying differential gene expression

Filters can contain DNA from a variety of organisms in the form of plasmid, cosmid, PAC, BAC, and YAC DNA, as well as PCR products. Currently, there are a variety of different DNA and cDNA membranes available. It has to be decided what the goal of the study is (e.g. is the aim to identify new genes, or to study the expression level of known genes in a particular setting) and how much is going to be spent. The most inexpensive membranes to date are colony arrays such as those from Genome Systems (`www.genomesystems.com`) and from the German Resource Center Primary Database (`www.rzpd.de`). The arrays of Genome Systems contain 18 432 double-spotted IMAGE clones on a 22 × 22 cm membrane. Arrays of mouse and yeast sequences are also available. The German Human Genome Project (`www.rzpd.de`) offers similar arrays. Both of these resources spot clones from the IMAGE consortium (`www-bio.llnl.gov/bbrp/image/image.html`) (22) on membranes, 80% of which have been partially sequenced through the Washington University Expressed Sequence Tag (EST) project. Alternatively, the GeneDiscoveryArray (GDA) from Genome Systems contains clones of 18 000 representatives of 'clusters' of ESTs that appear to be from the same gene determined by the UniGene database (`http://www.ncbi.nlm.nih.gov/UniGene/index.html`). The GDA costs about three times more than the standard array. Another option is to order a custom-made colony array (22 cm × 22 cm), which will contain a library of your choice (Genome Systems).

A disadvantage of the colony arrays is background hybridization due to host genome DNA. Arrays containing PCR products only are also available. Research Genetics, for example, offers 5 × 7 cm membranes that contain 50% PCR products from IMAGE clones of known human genes and 50% PCR products from IMAGE clones of genes with unknown function (Release I). Release II consists of PCR products of 1100 known genes. More releases are planned until 40 000 genes are represented. Finally, there are arrays where only known genes are represented by PCR products. For example, one can assess the role of 600 different genes that are known to be involved in apoptosis, cell cycle, transformation, cell growth, etc. under certain conditions using arrays of PCR products provided by Clontech. Because these membranes contain PCR products rather than clones from colonies, the portion of background due to hybridization to bacterial DNA will not occur. It may be useful to consult the above mentioned web pages for the expected updates, as the numbers and types of arrays available are constantly increasing.

4 Hybridization to the array

In our example, we hybridized [α-^{32}P]-labelled fingerprints from *Figure 1* to a colony array containing mouse IMAGE clones (Genome Systems) (see *Figures 2* and *3*). The proportion of genes detectable by a probe may improve for arrays of PCR products because the background hybridization is expected to be lower. The advantage of two-colour fluorescence is that two dyes can be used simultaneously on one array to detect differential gene expression, which controls internally for variability in the amount of spotted target sequence. Currently, fluorescence detection on nylon membranes is problematic due to background fluorescence from the membrane itself. We present a protocol for labelling with [α-^{32}P]dCTP. As we want to monitor differential gene expression after treatment with a chemopreventive drug, we will need two membranes, one for the untreated control and one for the probe following treatment of the cells. Two additional membranes can be addressed with the fingerprint probes generated from the RNA samples at lower concentrations to control for differences due to RNA concentration.

Background hybridization and incomplete blockage of Alu repeats limits the specificity of the hybridization signal. In our experience, with each arbitrarily primed probe derived from a cell line, we can reproducibly detect a different set

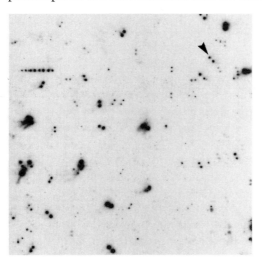

Figure 2 Differential hybridization to clone arrays: control. The image shows a close-up of an autoradiogram for a part of a larger membrane. It spans about 4000 double-spotted *E. coli* colonies, each carrying a different EST clone. RNA was prepared from untreated mouse thymocytes. The hybridization of a probe generated from the fingerprint of lane 1 of *Figure 1* is shown. The pair of radiolabelled colonies from one differentially expressed cDNA clone is indicated with an *arrow*.

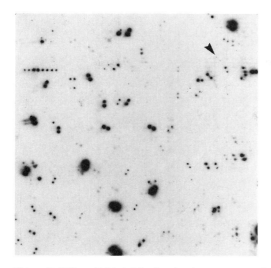

Figure 3 Differential hybridization to clone arrays: cells treated with sulindac. The image shows the same part as in *Figure 2* of a second membrane. Mouse thymocytes were treated with sulindac, a chemopreventive agent, prior to harvesting and RNA preparation. The hybridization of lane 3 of *Figure 1* is shown. The pair of radiolabelled colonies from one differentially expressed cDNA clone is indicated with an *arrow*. This transcript is down-regulated after sulindac treatment.

of 1000 of the 18 432 clones. We have shown that two different probes generated using different pairs of arbitrary primers will hybridize to largely non-overlapping sets of cDNAs in the same array (18) with fewer than 100 products overlapping among the most intensely hybridizing 2000 colonies. When comparing the hybridization of a RAP-PCR probe to that of a directly labelled mRNA population we found less than 5% overlap (18).

Protocol 5

Hybridization to the array

Equipment and reagents

- Hybridization oven with roller bottles (e.g. Techne HB-1D, VWR), or sealable plastic bags and a water-bath, set to 45°C
- Horizontal shaker
- Flat-bottomed glass or plastic container larger than the membranes to be used
- Table-top microcentrifuge (10 000 g)
- PhosphoImager (e.g. Molecular Dynamics) or X-ray film
- 20 × SSC: 3 M NaCl, 0.3 M Na$_3$ citrate.2H$_2$O pH 7.0
- 10% SDS

- 10 mg/ml fragmented, denatured salmon sperm DNA (Pharmacia)
- 50 × Denhardt's solution: 5 g Ficoll, 5 g polyvinylpyrrolidone, 5 g bovine serum albumin in 500 ml water and filter sterilized
- Formamide (Aldrich)
- 1 μg/μl fragmented human genomic DNA
- Poly(dA) oligonucleotide (20–25-mer) (Genosys)
- cDNA arrays (e.g. arrays purchased from Genome Systems)

Protocol 5 continued

Method

1. Pre-wash cDNA filters in three changes of 2 × SSC/0.1% SDS in a horizontally shaking, flat-bottomed container to reduce residual bacterial debris. Carry out the first wash in 500 ml for 10 min at room temperature, and the second and third washes in 1 litre of pre-warmed (50°C) solution for 10 min each at room temperature.

2. Pre-hybridize the filters in roller bottles in 60 ml pre-warmed (45°C) pre-hybridization solution containing 6 × SSC, 5 × Denhardt's reagent, 0.5% SDS, 100 μg/ml fragmented, denatured salmon sperm DNA, and 50% formamide for 1–2 h at 42–45°C.

3. Exchange the pre-hybridization solution with 10 ml pre-warmed (45°C) hybridization solution containing 6 × SSC, 0.5% SDS, 100 μg/ml fragmented, denatured salmon sperm DNA, and 50% formamide.

4. Denature sheared human genomic DNA mixed with 200 ng poly(dA) oligonucleotide in a boiling water-bath for 10 min. This procedure is used to decrease background hybridization due to repeats (e.g. Alu, Line elements, centromeric repeats). Immediately add to the hybridization solution to obtain a final concentration of 10 μg/ml of genomic DNA and 20 ng/ml of poly(dA) oligonucleotide.

5. At the same time, denature the labelled probe in a total volume of 280 μl in a boiling water-bath for 4 min, and immediately add to the hybridization solution.

6. Hybridize at 42–45°C for 12–48 h, typically 18 h.

7. Set the incubator oven temperature to 65°C. Pour off the hybridization solution, collect and dispose of it appropriately. Wash the membrane twice with 50 ml of 2 × SSC/0.1% SDS for 5 min. The wash solution is then replaced with 100 ml of 0.1 × SSC/0.1% SDS and incubated for 10 min.

8. Pre-warm wash solution (0.1 × SSC/0.1% SDS) to 68°C and wash for 40 min in 100 ml in the roller bottles.

9. Transfer the filter to a horizontally shaking, flat-bottomed container and wash in 1 litre pre-warmed (68°C) wash solution (0.1 × SSC/0.1% SDS) for 20 min under gentle agitation. Transfer filter back to a roller bottle containing 100 ml pre-warmed (68°C) 0.1 × SSC/0.1% SDS and incubate for 1 h. Remove the final wash solution and rinse the filter briefly in 2 × SSC at room temperature. Depending on the amount of radioactivity used, the wash solutions may be discarded.

10. Blot the membranes with 3MM paper and seal the moist membranes in moisture-proof bag. Expose membrane to phosphor screens or conventional X-ray film using intensifying screens.

11. Scan phosphor screens after 24 h exposure or develop films after 48 h.

5 Analysing arrays

To collect data, we expose the filter to a phosphor screen and scan it after an overnight exposure. We are currently using a Molecular Dynamics Storm 820 PhosphoImager system with *ArrayVision* software (Imaging Research Inc., also available from Genome Systems) to facilitate the gathering and processing of data. In this chapter, we have presented hybridization experiments with IMAGE clone colony arrays from Genome Systems, although the *ArrayVision* is originally designed for the analysis of GDA filters from Genome Systems. But because the membranes are of similar size (22 cm × 22 cm) and spotted in an identical fashion, it is possible to use the GDA grid for the ordinary EST colony filters as well. Background subtractions as well as normalization to a reference can be performed automatically after data collection. Also, *ArrayVision* allows multiple filters to be compared. Using *ArrayVision*, we have determined that the standard error between the intensities of two corresponding dots is about 10%. Higher errors, occurring in a few spots, are usually due to signal interference from adjacent bright spots due to the density of the grid.

6 Confirmation of differential gene expression using low stringency RT-PCR

To obtain the sequence of a candidate fragment from a colony array, cloning and sequencing is usually not necessary as 80% of the clones from the IMAGE consortium have been partially sequenced and deposited in the EST database of GenBank. All the clones are available (`www-bio.llnl.gov`) and can sequenced, if the sequence is unknown, or used for probe generation. Once the sequence of the gene is known, one can search for homologies to known genes using the *BLAST* algorithm at the NCBI web site (`http://www.ncbi.nlm.nih.gov`). To further identify the gene in question, the UniGene database (`http://www.ncbi.nlm.nih.gov/UniGene/index.html`), which presents clusters of human and mouse ESTs that appear to be from the same gene, is also useful. On the same web site, many of the UniGene clusters can be mapped onto chromosomes.

So far, cDNA libraries that contain near full-length clones are not yet available on arrays. 5′ RACE (23) can be used to obtain additional sequence information. If one uses arrays with PCR products that represent known genes (Clontech), cloning, sequencing, and sequence extension is, of course, redundant.

After identifying a potentially differentially expressed candidate gene, differential expression must usually be confirmed. The use of two RNA concentrations per sample in the differential hybridization experiments provides additional support for differential regulation and is more convenient when high throughput is required and the occasional error can be tolerated. Only the clones that show differential expression in the hybridization experiments with *both* RNA concentrations in both RNA samples (e.g. treated versus untreated) are considered. Ultimately, confirmation of differential expression of a gene can be done using RT-PCR of the known region of the expressed gene. For the confirmation of differential expression of a gene, a full-length sequence is not required and can be done using RT-PCR or other standard methods (e.g. Northern or RNase protection assay).

We use low stringency PCR to generate products of a few hundred bases in length (see *Protocol 6*). The internal 'control' PCR products that are produced because of the low stringency can be used to monitor the quality of the PCR reaction and the quality and quantity of the RNA used (17).

Following the confirmation of differential expression, one has to decide which of the target genes merit further studies. Genes that are already known might have the advantage that model systems are often available. Also, a known gene sometimes implies the involvement of other genes which can be analysed in the same experimental context. If the target gene is not known, one should search for homologies to known genes before proceeding with more involved functional studies, as the newly identified gene could be a family member of a known gene with known function.

Protocol 6
Low stringency RT-PCR

Equipment and reagents

- Thermocycler (Perkin Elmer Cetus)
- Sequencing size gel electrophoresis apparatus
- BioMax film (Eastman Kodak)
- Oligo(dT) primer or a set of random 9-mer primers (Genosys): 100 μM stock
- Pair of fragment-specific primers in 100 μM stock solution (Genosys)
- 2 × RT mix: 100 mM Tris pH 8.3, 150 mM KCl, 6 mM MgCl$_2$, 40 mM DTT, 0.4 mM of each dNTP, 1 μM of primer (either oligo(dT) or arbitrary primer), and 4 U/μl of M-MLV reverse transcriptase (Promega)
- Urea

- 2 × PCR mixture: 20 mM Tris pH 8.3, 20 mM KCl, 8 mM MgCl$_2$, 0.4 mM of each dNTP, 3 μM of each specific primer, 0.1 μCi/μl [α-^{32}P]dCTP (ICN), and 0.25 U/μl AmpliTaq® DNA polymerase Stoffel fragment (Perkin Elmer Cetus)
- Formamide/EDTA/XC/BPB gel loading buffer: 10 ml formamide (Aldrich), 10 mg xylene cyanol FF, 10 mg bromophenol blue, 200 μl 0.5 M EDTA pH 8.3
- 40% acrylamide stock solution (FisherBiotech)
- 10 × TBE pH 8.3 (see *Protocol 1*)

Method

1. Perform RT following *Protocol 2*, steps 1–4, using either an oligo(dT) primer or a set of random 9-mer primers.

2. Use 1.5 μM of each of the two specific primers (18–25-mers) in the PCR. Perform thermocycling similar to RAP-PCR (see *Protocol 2*, step 6) using 94°C for 1 min, 35–45°C for 1 min, and 72°C for 2 min. Cycle for 19, 22, and 25 cycles to control for variability in the abundance of transcripts, performance of primer pairs, and the amplifiability of the PCR products.

Protocol 6 continued

3. Mix an aliquot of the amplification products (3.5 μl) with 9 μl of formamide dye solution, denature at 85 °C for 4 min, and chill on ice. Load 2.4 μl onto a 5% polyacrylamide, 43% urea gel, prepared with 1 × TBE buffer. The PCR products resulting from the two different concentrations of the same RNA template are loaded side-by-side on the gel.

4. Perform electrophoresis at 1700 V or at a constant power of 50–70 W until the xylene cyanol tracking dye reaches the bottom of the gel (approx. 4 h). Dry the gel under vacuum and expose to a Kodak BioMax X-ray film for 16–48 h.

If one is confronted with too many potential target genes to choose from, it could be valuable to attempt to induce fewer differentially expressed genes. In our example of treating with a chemopreventive drug, this could mean decreasing the duration and/or dose of the drug treatment to find transcripts that respond early, or that are more sensitive to the treatment. A more complex approach is to prepare multiple probes from cells treated with different agents in order to further narrow down the phenomenon of interest (14, 19). In our example, as we are interested in programmed cell death induced by our chemo-preventive treatment, we can compare the data after drug-induced apoptosis to the data obtained from cells that have been stressed, but which do not undergo apoptosis. Then, the differentially expressed genes can be divided into apoptosis responses, stress responses, and a large class of genes that respond to both.

It would be very useful to have public resources that link every gene to the transcriptional effect in a growing list of conditions. Technologies such as the one we are presenting in this chapter will help generate the data that allow such resources to develop. The knowledge of promoters, transcriptional targets, and pathways of the gene in question would greatly facilitate further functional studies of newly identified differentially expressed target genes and help to better and more quickly understand pathways in molecular biology.

References

1. Pietu, G., Alibert, O., Guichard, V., Lamy, B., Bois, F., Leroy, E., *et al.* (1996). *Genome Res.*, **6**, 492.
2. Ramsay, G. (1998). *Nature Biotechnol.*, **16**, 40.
3. Marshall, A. and Hodgson, J. (1998). *Nature Biotechnol.*, **16**, 27.
4. DeRisi, J., Penland, L., Brown, P. O., Bittner, M. L., Meltzer, P. S., Ray, M., *et al.* (1996). *Nature Genet.*, **14**, 457.
5. Schena, M., Shalon, D., Davis, R. W., and Brown, P. O. (1995). *Science*, **270**, 467.
6. Schena, M., Shalon, D., Heller, R., Chai, A., Brown, P. O., and Davis, R. W. (1996). *Proc. Natl. Acad. Sci. USA*, **93**, 10614.
7. Chee, M., Yang, R., Hubbell, E., Berno, A., Huang, X. C., Stern, D., *et al.* (1996). *Science*, **274**, 610.
8. Weiler, J., Gausepohl, H., Hauser, N., Jensen, O. N., and Hoheisel, J. D. (1997). *Nucleic Acids Res.*, **25**, 2792.
9. Rhyner, T. A., Biguet, N. F., Berrard, S., Borbely, A. A., and Mallet, J. (1986). *J. Neurosci. Res.*, **16**, 167.

10. Pesole, G., Liuni, S., Grillo, G., Belichard, P., Trenkle, T., Welsh, J., *et al.* (1998). *BioTechniques*, **25**, 112.
11. Welsh, J. and McClelland, M. (1990). *Nucleic Acids Res.*, **18**, 7213.
12. Welsh, J., Chada, K., Dalal, S. S., Cheng, R., Ralph, D., and McClelland, M. (1992). *Nucleic Acids Res.*, **20**, 4965.
13. Ralph, D., McClelland, M., and Welsh, J. (1993). *Proc. Natl. Acad. Sci. USA*, **90**, 10710.
14. McClelland, M., Ralph, D., Cheng, R., and Welsh, J. (1994). *Nucleic Acids Res.*, **22**, 4419.
15. Vogt, T. M., Welsh, J., Stolz, W., Kullmann, F., Jung, B., Landthaler, M., *et al.* (1997). *Cancer Res.*, **57**, 3554.
16. Vogt, T., Stolz, W., Welsh, J., Jung, B., Kerbel, R. S., Kobayashi, H., *et al.* (1998). *Clin. Cancer Res.*, **4**, 791.
17. Jung, B., Vogt, T., Mathieu-Daude, F., Welsh, J., McClelland, M., Trenkle, T., *et al.* (1998). *Carcinogenesis*, **19**, 1901.
18. Trenkle, T., Welsh, J., Mathieu-Daude, F., Jung, B., and McClelland, M. (1998). *Nucleic Acids Res.*, **26**, 3883.
19. McClelland, M., Mathieu-Daude, F., and Welsh, J. (1995). *Trends Genet.*, **11**, 242.
20. Liang, P. and Pardee, A. B. (1992). *Science*, **257**, 967.
21. Kebelmann-Betzig, C., Seeger, K., Dragon, S., Schmitt, G., Moricke, A., Schild, T. A., *et al.* (1998). *BioTechniques*, **24**, 154.
22. Lennon, G. G., Auffray, C., Polymoropoulos, M., and Soares, M. B. (1996). *Genomics*, **33**, 151.
23. Zhang, Y. and Frohman, M. A. (1997). *Methods Mol. Biol.*, **69**, 61.

Chapter 5

The use of RT-PCR differential display in single-celled organisms and plant tissues

MICHAEL L. NUCCIO, TZUNG-FU HSIEH, and
TERRY L. THOMAS

Department of Biology, Texas A&M University, College Station,
Texas 77843-3258, U.S.A.

1 Introduction

Differential display of eukaryotic mRNA is a fast and powerful PCR-based technique developed to compare relative variation in gene expression between RNA populations (1, 2). This technique provides a way to monitor differentially expressed genes across various tissues, developmental stages, or physiological states by systematically amplifying cDNA derived from mRNA populations and displaying them on a polyacrylamide gel. The bands that are identified as potential targets of interest can subsequently be isolated, reamplified, and cloned for further verification. Differential display is flexible in two ways. First, it is unnecessary to begin experiments with a poly(A) enriched RNA fraction; they can be performed with as little as 1 μg of total RNA per sample. Secondly, many comparisons between samples can be made using various primer combinations in the assay.

Despite the wide use of differential display for gene expression analysis, several problems remain associated with the technique (3). One problem intrinsic to the method is the high frequency of 'false positives' among the cloned PCR fragments which cannot be verified by RNA gel blot analysis or other methods. Moreover, the fact that a single differential display band may contain more than one cDNA fragment of similar length further complicates downstream analysis. Several refinements and optimizations have since been made to help reduce the effort and streamline the verification process (2–6). Systematic comparisons have been conducted among different subtraction methods, and differential display was concluded to be the method of choice because it identifies mRNAs independent of prevalence (7).

This chapter outlines procedures that have been successfully applied in plants. Emphasis will be placed on RNA preparation and verification of differential display products. The application of the differential display in single-celled organisms and fungi has also been reported (8–12) and will not be reviewed here. The approaches are similar to those presented in this chapter.

2 Suitable total RNA preparation procedures

Any protocol that is capable of isolating clean and undegraded total RNA can be used with differential display analysis. Many RNA isolation protocols have been reported with success. In our experience, RNAzol (Tel-Test, Inc.) yielded good total RNA suitable for differential display. Tri-Reagent (Molecular Research Center) is another total RNA isolation reagent that produces acceptable RNA in a relatively short time and has been used successfully in differential display experiments (13). Others have used either phenol/SDS (14) or guanidinium isothiocyanate/ phenol/ chloroform (15) based methods to isolate plant total RNAs for differential display experiments (16, 17).

Many RNA preparation kits are commercially available. Often these kits greatly simplify the procedure of preparing high quality RNA. However, the suitability of the RNA prepared with these kits for differential display analysis remains a matter of empirical analysis. The quality of RNA isolated using any method should be verified by gel electrophoresis prior to first strand cDNA synthesis.

The quality of the starting material for differential display is critical. For this procedure, RNA must be DNA-free and be intact as determined by an RNA gel profile. There are many protocols available for total RNA preparation, and we recommend using methods that were developed for the target sample(s) or similar tissues. Many of these procedures can be scaled down. Typically the differential display procedure requires 200 mg to 1 g fresh weight of starting plant material. The total RNA preparation outlined in *Protocol 1* has been modified from Crouch *et al.* (18) as described by Nuccio and Thomas (19). This protocol, though lengthy, allows preparation of clean total RNA from mature seeds and other recalcitrant tissues. General precautions for RNA handling should be followed (14).

Protocol 1

Preparation of total RNA using hot phenol method[a]

Equipment and reagents

- Polytron
- HB buffer: 0.1 M Tris–HCl pH 9.0, 0.1 M NaCl, 1 mM EDTA pH 8.0, 0.5% SDS
- Chloroform:isoamyl alcohol (24:1)
- 5 M KOAc pH 6.0
- 4 M LiCl

Method

1. Grind tissue to a fine powder in liquid nitrogen with a mortar and pestle.

2. In a 50 ml OakRidge tube at 4°C, resuspend powder in 20 ml/g tissue (v/w) homogenization buffer. Vortex until no clumps are visible.

3. Add one-half volume of hot HB-saturated phenol (60°C). Homogenize with a Polytron for 1 min at high speed.

4. Add one-half volume chloroform:isoamyl alcohol (24:1) and homogenize as in step 3.

5. Separate phases by centrifugation at 8000 g for 10 min at 4°C.

6. Recover aqueous (upper) phase and repeat steps 3–5.

7. Recover aqueous phase in a 30 ml Corex tube, add 5 M KOAc to 0.2 M, then add 2 vol. of 100% ethanol.

8. Leave at −20°C overnight.

9. Collect precipitated nucleic acids at 12 000 g for 15 min at 4°C. Wash precipitated nucleic acid with 70% ethanol. Dry pellet under vacuum.

10. Dissolve precipitate in 400 μl TE and transfer to a 1.5 ml Eppendorf tube. Repeat steps 7–9.

11. Add an equal volume of 4 M LiCl. Leave overnight at 0°C. Keep the tube on ice in a cold room if possible.

12. Pellet RNA in a microcentrifuge at full speed for 15 min at 4°C. Remove as much supernatant as possible. Dissolve RNA in 400 μl TE and add 1.5 vol. of 5 M KOAc. Leave at −20°C for 5 h.

13. Pellet RNA as in step 12. Resuspend pellet in 400 μl TE, add 5 M KOAc to 0.2 M, and add 2 vol. of 100% ethanol. Leave overnight at −20°C.

14. Recover RNA as in step 9. Dissolve RNA in a small volume of TE. Determine the concentration by UV absorption. The A_{260}/A_{280} ratio should be close to 2.0. Check RNA quality by resolving 1 μg on a small formaldehyde agarose gel.

15. For long-term storage, keep RNA samples at −80°C as ethanol precipitates until needed.

[a] Modified from the method given in ref. 18.

3 Differential display analysis

Differential display can provide a resource of putative tissue-, temporal-, or stage-specific probes. It is extremely sensitive, opening up a variety of opportunities previously unattainable through conventional molecular analysis. This method is quick, yielding results within a matter of days. Furthermore, if an amplification product can be detected, it can be cloned or utilized as a probe for subsequent characterization or isolation of the corresponding full-length cDNA clone. It is the downstream analysis of target bands that reveals the considerable background associated with this technology. It is noteworthy that a single differential display band can contain a heterogeneous mixture of PCR reaction products (5, 19). This requires some effort to resolve. Section 4 outlines a variety of strategies that have been used to successfully identify bona fide differentially expressed cDNAs.

In practice differential display requires up-front control to ensure fidelity and reproducibility. The PCR amplification step is very sensitive and is prone to experimental fluctuations. To aid in reducing this background we, and others, have found that it is necessary to make two RNA preparations for each sample and synthesize first strand cDNA templates individually. Each cDNA template should then be used in duplicate PCR reactions to produce a total of four differential display profiles per sample. Only those signals which are reproducible should be targeted for further characterization.

This section presents the methodology for differential display analysis. There are three primary components. The first strand cDNA synthesis is outlined first. This is followed by two methods for the differential display PCR reaction. The first method (*Protocol 3*) is essentially that of Liang and Pardee (1). The second method (*Protocol 5*) is derived from the first one with modification. The only difference being that an end-labelled, anchored primer was used in the second method instead of $[^{32}P]dATP$ to identify the amplified products. In designing a differential display experiment take note that some primer combinations yield better results than others. Therefore, several candidate bands should be generated with multiple primer pairs to produce as many target signals as possible and thus to maximize the number of novel genes identified.

3.1 First strand cDNA synthesis

In our experiments we have empirically determined that 2 µg total RNA per sample is the benchmark for first strand cDNA synthesis. The amount depends on the differential display PCR method that is chosen. *Protocol 3* is effective with as little as 20 ng of total RNA. However *Protocol 5* loses sensitivity if less than 1 µg total RNA is used. There are a variety of cDNA synthesis kits available commercially. In our experiments reagents from Gibco BRL's cDNA synthesis system were employed.

25 µM of $T_{11\ or\ 12}VN$ (V represents A, C, or G; N represents all four nucleotides) primer stocks were made up in DEPC H_2O. These stocks were used in both first strand cDNA synthesis and differential display reactions.

Protocol 2

First strand cDNA synthesis

Equipment and reagents

- cDNA synthesis system (Gibco BRL)
- DEPC treated water
- 25 µM $T_{11\ or\ 12}VN$ oligonucleotide primer

Method

1. Mix the following reagents in a microcentrifuge tube:
   - 1 µl total RNA (2 µg)
   - 5 µl of 5 × reverse transcriptase reaction buffer

Protocol 2 continued

- 2.5 µl of 25 µM $T_{11 \text{ or } 12}VN$ primer
- 2.5 µl of 200 µM dNTP mix
- 11.5 µl DEPC H_2O

2. Mix by pipetting gently, then heat to 65°C for 3 min, then let stand at room temperature for 3 min. Repeat this step one more time.
3. Allow the reaction mix to cool to room temperature then add:
   - 1.25 µl of 0.1 M DTT
   - 1.25 µl Superscript (250 U)
4. Mix by pipetting.
5. Incubate at 42°C for 60 min.
6. Heat inactivate reaction by incubating at 95°C for 5 min.

The first strand cDNA synthesis is now complete and samples can be stored at −20°C. They should be stable for about two months.

3.2 Differential display PCR amplification

The experimental procedures described in this section basically followed the original protocol published by Liang and Pardee (1) with minor modifications. The reaction products are labelled with $[^{32}P]dATP$ instead of $[^{35}S]dATP$. Also the amount of cDNA template is slightly higher. This protocol has been used successfully by many researchers (20, 21). However it does generate a very low signal-to-noise ratio. As discussed earlier, this confounds target band isolation. In Section 3.3 we present a series of modifications to this protocol which greatly improve the reproducibility.

The following procedure utilizes cDNA templates generated from *Protocol 2*. If several reactions are to be done, the components should be made up as a master mix. This will aid in maintaining uniformity within the experiment. Make enough cocktail for all the samples plus one to account for pipetting error. Since $[^{32}P]dATP$ is used instead of $[^{35}S]dATP$ the band regeneration step (*Protocols 7 or 8*) should immediately follow to avoid the possibility of radiolysis.

Protocol 3

Differential display PCR amplification using internal labelling

Reagents

- First strand cDNA templates (products from *Protocol 2*)
- 25 µM $T_{11 \text{ or } 12}VN$ oligonucleotide primer (see *Protocol 2*)
- 5 µM arbitrary 10-mer
- *Taq* polymerase and buffer (Promega)
- 200 µM dNTP mix
- $[\alpha^{-32}P]dATP$, 800 Ci/mmol (Du Pont)
- Sequencing stop buffer (Epicenter Technologies)

Protocol 3 continued

Method

1. Mix the following components in a 0.5 ml microcentrifuge tube:

   - 2.0 μl of first strand cDNA template
   - 2.5 μl of 25 μM $T_{11 \text{ or } 12}VN$
   - 2.5 μl of arbitrary 10-mer
   - 2.5 μl of 10 × *Taq* polymerase buffer
   - 2.5 μl of dNTP mix
   - 1.0 μl of $[^{32}P]$dATP
   - 1.5 μl of 25 mM $MgCl_2$
   - 9.5 μl of MQ H_2O
   - 1.0 μl of *Taq* DNA polymerase (5 U)

   Overlay the reaction contents with mineral oil, if necessary.

2. Perform PCR at the following cycling conditions: 94°C for 5 min; 40 cycles of 94°C for 30 sec, 42°C for 60 sec, 72°C for 30 sec; and then 72°C for 5 min.

3. Resolve amplification products on a 6% sequencing gel. Combine 5 μl of each PCR amplification product with 3 μl sequencing stop buffer and heat to 94°C for 5 min. Load samples and run the gel until the xanthine cyanol (the light blue dye) is half-way down the gel. This will resolve reaction products ranging in size from 200–500 bases. A ^{32}P-labelled DNA ladder can be used for sizing purposes.

4. Dry gel on 3MM Whatman paper for support and expose to X-ray film at −80°C. With these conditions exposure times will vary from several minutes to a few hours.[a]

[a] Alternative methods to detect differential display bands such as a PhosphoImager may be used, however, the resulting two-dimensional output is not identical in size to the original gel. This may make gel alignment and band isolation more problematic.

3.3 Modifications to the PCR reaction

Many differential display bands identified using *Protocol 3* were flanked by the arbitrary 10-mer only. This is not limited to the primer sequence. Due to this observation *Protocol 3* was modified so that differential display identifies PCR products by end-labelled, anchored primer rather than by internal-labelling during amplification (*Protocol 5*). This significantly decreased the number of bands displayed on the gel. Experiments with this method also demonstrated that under the conditions in *Protocol 3*, the $T_{11 \text{ or } 12}VN$ primed very inefficiently. As is shown in *Figure 1*, the $T_{11 \text{ or } 12}VN$ primer performs poorly at a 42°C annealing temperature but begins to function at temperatures closer to 35°C. A simple calculation reveals that the T_m of each primer used in these reactions ranges

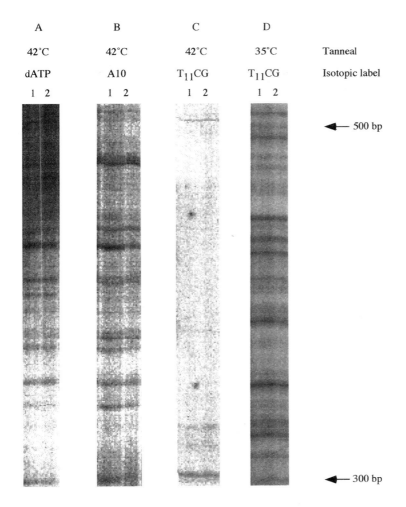

Figure 1 Primer behaviour in the differential display PCR reaction. Differential display reactions were performed on *Arabidopsis* leaf tissue (var. Landsberg) to determine the efficiency of each primer in the differential display PCR amplification. The cDNA was synthesized from 1 μg total RNA using the 5'-T_{11}CG-3' primer. cDNA representing 40 ng of the total RNA template was used in differential display PCR amplifications that were primed with the 5'-T_{11}CG-3' and the arbitrary decamer A10 (5'-GTGACTGCAG-3') oligonucleotides. The reaction products were labelled as follows: (A) [^{32}P]dATP with a 42°C annealing temperature, (B) ^{32}P-labelled A10 decamer with a 42°C annealing temperature, (C) ^{32}P-labelled 5'-T_{11}CG-3' with a 42°C annealing temperature, and (D) ^{32}P-labelled 5'-T_{11}CG-3' with a 35°C annealing temperature. Figure and legend reproduced from *Figure 1* in ref. 21 with kind permission from Kluwer Academic Publishers.

over several degrees but the 10-mer has a T_m (30°C–40°C) that is significantly higher than the T_m of $T_{11\ or\ 12}$VN (26°C–30°C).

This may explain why most of the differential display products identified using *Protocol 3* were flanked by the 10-mer rather than by both primers. By using labelled $T_{11 \text{ or } 12}$VN primer, we are essentially selecting for reaction products that contain the $T_{11 \text{ or } 12}$VN primer and thus greatly reduce the background generated by the arbitrary 10-mer. The noise increased due to reducing the annealing temperature in the PCR is not apparent because it is not labelled. *Protocol 4* describes procedures for end-labelling oligonucleotide primers.

Protocol 4

Labelling the $T_{11 \text{ or } 12}$VN primer

This reaction will produce enough labelled $T_{11 \text{ or } 12}$VN primer for about 25 differential display reactions.

Reagents

- $[\gamma\text{-}^{32}P]$ATP, 6000 Ci/mmol (Du Pont)
- 95 mM of $T_{11 \text{ or } 12}$VN primer
- T4 polynucleotide kinase and buffer (NEB)
- 1 mg/ml BSA
- 0.5 M EDTA pH 8.0
- 10 mg/ml glycogen (Sigma)
- 7.5 M NH_4OAc

Method

1. Mix the following components in a microcentrifuge tube:
   - 33 µl of 95 mM $T_{11 \text{ or } 12}$VN primer
   - 5 µl of $[\gamma\text{-}^{32}P]$ATP (6000 Ci/mmol)
   - 2 µl of BSA
   - 5 µl of $10 \times$ kinase buffer
   - 2 µl T4 polynucleotide kinase (10 U)
   - 3 µl MQ H_2O

2. Incubate at 37°C for 1 h.

3. Terminate the reaction by adding 5 µl of 0.5 M EDTA pH 8.0.

4. Precipitate the labelled primer by adding 50 µg glycogen, 27.5 µl of 7.5 M NH_4OAc, and 137.5 µl of ice-cold 95% ethanol. Let the precipitation stand at −80°C overnight.

5. Recover DNA by centrifugation at full speed for 30 min at 4°C. Wash labelled primers with 50 µl of 95% ethanol and dry in a SpeedVac.

6. Dissolve primer in 51 µl TE and determine ^{32}P incorporation by counting 0.5 µl. Incorporation should be $5.0\text{–}8.0 \times 10^8$ c.p.m.

7. Store at −20°C until use.

Protocol 5 is essentially the same as described in *Protocol 3*. The only changes are the radioactive label, dNTP concentration and the annealing temperature during PCR. Each reaction includes $1.0\text{–}2.0 \times 10^7$ c.p.m. of labelled $T_{11 \text{ or } 12}$VN primer.

Protocol 5

Differential display PCR amplification using end-labelled, anchored primer

Reagents

- First strand cDNA templates (products from *Protocol 2*)
- End-labelled $T_{11 \text{ or } 12}VN$ oligonucleotide primer (see *Protocol 4*)
- Arbitrary 10-mer (see *Protocol 3*)
- *Taq* polymerase and buffer (Promega)
- 2 mM dNTP mix
- Sequencing stop buffer (Epicenter Technologies)

Method

1. Mix the following components in a 0.5 ml microcentrifuge tube:
   - 2.0 μl of first strand cDNA template
   - 2.5 μl of $T_{11 \text{ or } 12}VN$
   - 2.5 μl of arbitrary 10-mer
   - 2.5 μl of 10 × *Taq* polymerase buffer
   - 2.5 μl of dNTP mix
   - 1.5 μl of 25 mM $MgCl_2$
   - 10.5 μl of MQ H_2O
   - 1.0 μl of *Taq* DNA polymerase (5 U)

   Overlay the reaction contents with mineral oil, if necessary.

2. Perform PCR at the following cycling conditions: 94°C for 5 min; 40 cycles of 94°C for 30 sec, 35°C for 60 sec, 72°C for 30 sec; and then 72°C for 5 min.

3. Resolve amplification products on a 6% sequencing gel. Combine 5 μl of each PCR amplification product with 3 μl sequencing stop buffer and heat to 94°C for 5 min. Load samples and run the gel until the xanthine cyanol (the light blue dye) is half-way down the gel. This will resolve reaction products ranging in size from 200–500 bases. A ^{32}P-labelled DNA ladder can be used for sizing purposes.

4. Dry gel on 3MM Whatman paper for support and expose to X-ray film at −80°C. With these conditions exposure times will vary from several minutes to a few hours.[a]

[a] Alternative methods to detect differential display bands such as PhosphoImager may be used, however, the resulting two-dimensional output is not identical in size to the original gel. This may make gel alignment and band isolation more problematic.

An example of differential display result using *Protocol 5* is shown in *Figure 2*. Two bands were identified as seed-specific genes. Both cDNA fragments were subsequently isolated and characterized using methods described in Section 4.

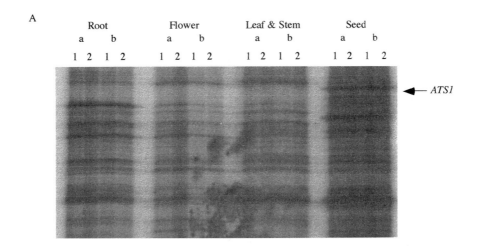

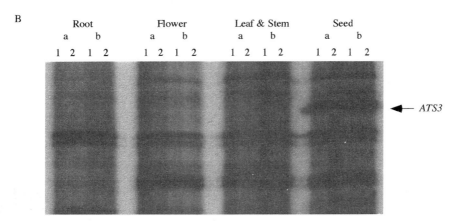

Figure 2 Identification of *ATS1* and *ATS3* by differential display. The differential display PCR amplifications were primed with the 5′-T$_{12}$CG-3′ and the (A) A10 (5′-GTGACTGCAG-3′), and (B) ca/b (5′-CTAGCTTGGT-3′) oligonucleotides. To assess reproducibility, the cDNA representing each sample was synthesized in two independent reactions (1 and 2), and the PCR amplification was performed twice on each cDNA template (a and b). Bands representing the *ATS1* and *ATS3* genes are indicated. Figure and legend reproduced from *Figure 1* in ref. 19 with kind permission from Kluwer Academic Publishers.

4 Band isolation, amplification, and downstream analysis

Once potential target bands have been identified they must be isolated. This should be done quickly if *Protocol 3* is used. It will be necessary to use radioactive ink to carefully mark the gel. A good autoradiograph is important in locating particular bands. DNA is recovered using a standard electroelution protocol (14).

A   B

ethidium stain   autoradiogram

100 bp   ddp1   ddp2   100 bp   ddp1 ddp2

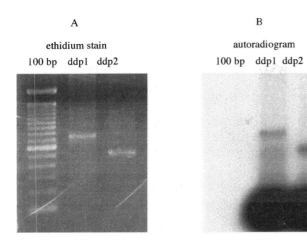

Figure 3 Regeneration of differential display products. The DNA representing target differential display products was excised from the dry sequencing gel, electroeluted, and precipitated. This DNA was used as template in a PCR amplification to regenerate the DNA template. The PCR amplifications were primed with the $5'$-T_{11}CG-$3'$ and the A10 $5'$-GTGACTGCAG-$3'$ oligonucleotides. Each reaction contained a small amount of ^{32}P-labelled $5'$-T_{11}CG-$3'$ to identify the reaction products on an autoradiogram. The annealing temperature for the PCR amplification was 35°C. The reaction products were resolved on a 1.5% agarose gel which was stained with ethidium bromide and photographed (A). The gel was dried down and autoradiographed (B).

The DNA amplification step incorporates a small amount of ^{32}P-labelled $T_{11\ or\ 12}$VN primer as an internal control. *Figure 3* is an example of a positive result. The amplification product should be of the expected size and contain a small amount of ^{32}P-label.

Protocol 6

Regenerating differential display bands

Reagents

- Glycogen (see *Protocol 4*)
- NH$_4$OAc (see *Protocol 4*)
- $T_{11\ or\ 12}$VN oligonucleotide primer (see *Protocol 4*)
- Arbitrary 10-mer (see *Protocol 3*)
- *Taq* polymerase and buffer (Promega)
- 2 mM dNTP mix

Method

1. Expose the gel long enough to get a good signal, carefully align the gel and the autoradiograph, and excise the band with a scalpel blade. To be certain that the appropriate band(s) were excised, re-expose the gel and the excised band(s) should have disappeared.

2. Place the gel slice (paper and gel) into 300 µl of 1 × TBE electrophoresis buffer in a dialysis bag (6000–8000 M$_r$ cut-off).

Protocol 6 continued

3. Electroelute the DNA at 25 mA for 2 h.

4. Reverse the electrodes and continue to elute for another 30 sec.

5. Carefully remove the eluent to a 1.5 ml centrifuge tube. Centrifuge for 5 min in a microcentrifuge at full speed to remove debris. Remove supernatant to a clean 1.5 ml centrifuge tube.

6. Precipitate the DNA by adding 2 μl glycogen as carrier, 0.5 vol. of 7.5 M NH_4OAc, and 2.5 vol. of ethanol. Precipitate overnight at $-20\,^{\circ}C$ or 1 h at $-80\,^{\circ}C$.

7. Centrifuge at full speed for 15 min at $4\,^{\circ}C$, wash precipitated DNA with 200 μl of 95% ethanol, and dry in a SpeedVac. Dissolve DNA in 10 μl TE. This DNA represents the differential display band. A portion of it will be used to amplify the DNA. Store the remaining DNA at $-20\,^{\circ}C$.

8. To reamplify the recovered differential display bands, mix the following components in a 0.5 ml microcentrifuge tube:
   - 4 μl recovered differential display band DNA
   - 2.5 μl of 25 μM $T_{11\ or\ 12}VN$ primer
   - 1 μl of ^{32}P-labelled $T_{11\ or\ 12}VN$ primer
   - 2.5 μl of 5 μM arbitrary 10-mer
   - 2.5 μl of $10 \times$ *Taq* polymerase buffer
   - 4 μl of 2.0 mM dNTP mix
   - 2 μl of 25 mM $MgCl_2$
   - 10.5 μl of MQ H_2O
   - 1.0 μl of *Taq* DNA polymerase (5 U)

9. Perform PCR using cycling program described in *Protocol 5*.

10. Precipitate the DNA by adding 2 μl glycogen, 0.5 vol. of NH_4OAc, and 2.5 vol. of ethanol. Incubate overnight at $-20\,^{\circ}C$ or 1 h at $-80\,^{\circ}C$.

11. Centrifuge at full speed for 15 min at $4\,^{\circ}C$, wash precipitated DNA with 200 μl of 95% ethanol, and dry in a SpeedVac. Resuspend in 20 μl TE. Store at $-20\,^{\circ}C$ until needed.

The amplified product should now be concentrated enough to be visible on a 1.5% agarose gel. This DNA can now be cloned and sequenced or used to generate DNA probes. The rest of the amplified product can be stored at $-80\,^{\circ}C$ for future use.

4.1 DNA cloning

The DNA representing differential display bands can be cloned into a blunt-ended plasmid vector such as *Sma*I digested pBluescript™ (Stratagene). There are also a variety of PCR product cloning systems available. It is important to note that a heterogeneous population of PCR fragments of similar size may be present in a single differential display band. This is where 'false positives' are often discovered as the differential display band may consist of the target sequence and a variety of additional amplification products. Unambiguous confirmation that a clone represents a differentially expressed gene requires verification by other methods. This can be done by either RNA gel blot analysis or RT-PCR with nested primers. The time required to identify true targets varies and may even be unsuccessful in some cases.

Two possible approaches to clone differential display products are presented. *Protocol 7* is based on a blunt-end ligation into a dephosphorylated vector, and *Protocol 8* incorporates unique restriction sites into the reamplified differential display band DNA which is then ligated into the appropriately digested plasmid DNA.

Protocol 7

Cloning differential display bands by blunt-end ligation

Reagents

- T4 DNA polymerase (NEB)
- dNTP mix (see *Protocol 5*)
- Glycogen (see *Protocol 4*)
- NH$_4$OAc (see *Protocol 4*)
- T4 DNA kinase (NEB)
- Shrimp alkaline phosphatase (Amersham)
- T4 DNA ligase (NEB)

Method

1. T4 DNA polymerase is used to polish the ends of the reamplified DNA. Take 15 μl of the reamplified DNA and add 2 μl of 10 × T4 DNA polymerase buffer, 1 μl dNTPs, and 5 U of T4 DNA polymerase in a 20 μl reaction. Incubate at 37 °C for 15 min.

2. Precipitate the DNA by adding 2 μl glycogen, 0.5 vol. of NH$_4$OAc, and 2.5 vol. of ethanol. Incubate overnight at −20 °C or 1 h at −80 °C.

3. Centrifuge at full speed for 15 min at 4 °C, wash precipitated DNA with 200 μl of 95% ethanol, and dry in a SpeedVac. Resuspend in 17 μl MQ H$_2$O.

4. T4 DNA kinase is used to phosphorylate the reamplified DNA. Add 2 μl of 10 × T4 DNA kinase buffer containing 10 mM ATP and 5 U of T4 DNA kinase. Incubate at 37 °C for 15 min. Precipitate DNA as in step 2. Resuspend in 4 μl MQ H$_2$O.

5. Prepare the cloning vector by digesting 5 μg vector DNA with *EcoRV* or *SmaI* in a 40 μl reaction. Incubate at 37 °C (25 °C for *SmaI*) for 2 h. Add 4.6 μl of 10 × shrimp alkaline phosphatase buffer and 1.5 U shrimp alkaline phosphatase. Incubate at 37 °C for an additional hour. Terminate the reaction at 65 °C for 15 min.

6. Precipitate the vector DNA by adding 2 μl glycogen, 0.5 vol. of NH$_4$OAc, and 2.5 vol. of ethanol. Incubate overnight at −20 °C or 1 h at −80 °C.

7. Centrifuge at high speed for 15 min at 4 °C, wash precipitated DNA with 200 μl of 95% ethanol, and dry in a SpeedVac. Dissolve DNA in 8 μl MQ H$_2$O.

8. To set up the ligation reaction, combine 4.0 μl differential display product, 4.0 μl vector, 1.0 μl of 10 × T4 DNA ligase buffer, and 1.0 μl T4 DNA ligase (400 U) for a 10 μl reaction. Set up a vector-alone control ligation using the remaining 4 μl vector DNA. Incubate overnight at 16°C.

9. Transform 1 μl of the ligation reaction into bacterial cells and plate onto selection medium.

Protocol 8 describes an alternative approach to clone the reamplified differential display products. This procedure requires that the differential display band DNA be reamplified with a set of modified oligonucleotide primers. They should be identical to those used to do the differential display PCR amplification except that they include restriction sites at their 5′ ends. A distinct restriction site with a six base recognition sequence is added 5′ to the primer sequence, i.e. an *Eco*RI site on the arbitrary 10-mer and a *Bam*HI site on the $T_{11 \text{ or } 12}$VN primer. Keep in mind that some restriction enzymes do not cut effectively near DNA ends.

Protocol 8

Cloning differential display bands by directional ligation

Reagents

- Modified $T_{11 \text{ or } 12}$VN primer and arbitrary 10-mer
- T4 DNA polymerase (NEB)
- dNTP mix (see *Protocol 5*)
- Glycogen (see *Protocol 4*)
- NH$_4$OAc (see *Protocol 4*)
- T4 DNA kinase (NEB)
- Shrimp alkaline phosphatase (Amersham)
- T4 DNA ligase (NEB)

Method

1. Perform *Protocol 6* except substitute the modified primers, with the unique restriction sites, for the primers used in the differential display PCR amplification. Dissolve the DNA in 11.5 μl MQ H$_2$O.

2. Set up a double digestion of the DNA in a 20 μl reaction. Incubate at 37°C for 2 h. Gel purify the digested DNA.

3. Precipitate the DNA by adding 2 μl glycogen, 0.5 vol. of NH$_4$OAc, and 2.5 vol. of ethanol. Incubate overnight at −20°C or 1 h at −80°C.

4. Centrifuge at full speed for 15 min at 4°C, wash DNA pellet with 200 μl of 95% ethanol, and dry in a SpeedVac. Resuspend in 4 μl MQ H$_2$O.

5. Prepare the cloning vector by digesting 5 μg vector DNA with the appropriate restriction enzymes in 40 μl. Incubate at 37°C for 2 h. Add 4.6 μl of 10 × shrimp alkaline phosphatase buffer and 1.5 U shrimp alkaline phosphatase. Incubate at 37°C for an additional hour. Terminate the reaction at 65°C for 15 min.

Protocol 8 continued

6. Precipitate the vector DNA by adding 2 μl glycogen, 0.5 vol. of NH$_4$OAc, and 2.5 vol. of ethanol. Incubate overnight at $-20\,°C$ or 1 h at $-80\,°C$.

7. Centrifuge at full speed for 15 min at 4°C, wash DNA pellet with 200 μl of 95% ethanol, and dry under vacuum. Resuspend in 8 μl MQ H$_2$O.

8. Set up the ligation by combining 4.0 μl differential display product, 4.0 μl vector, 1.0 μl of 10 × T4 DNA ligase buffer, and 1.0 μl T4 DNA ligase (400 U) for a 10 μl reaction. Set up a vector-alone control ligation using the remaining 4 μl vector DNA. Incubate overnight at 16°C.

9. Transform ligation into bacterial cells.

4.2 Generation and use of differential display probes

One of the problems with differential display is that it utilizes a total RNA template. Background associated with amplification from ribosomal RNAs is inevitable. Many have used DNA dot blot/slot blot analysis to preclude false positive clones (5, 6, 22). We have taken an alternative approach to use the reamplified differential display fragment to screen a target cDNA library directly. The reamplified differential display products are used as template to synthesize high specific-activity DNA probes to screen a cDNA library representing the target tissue. The cDNA library represents the poly(A)-enriched RNA fraction of the target tissue; spurious DNA amplified from non-poly(A) RNA will not hybridize. The second advantage of this approach is that full-length cDNAs representing target genes are identified. Plaque purified cDNAs rarely represent a single gene. In the experiment described in *Protocol 9* we recommend that at least 12 independent cDNAs be plaque purified. In our experience, a bona fide target gene will usually be represented in about half the cDNAs. Each unique cDNA should be examined by RNA gel blot analysis or RT-PCR to confirm its expression pattern.

Protocol 9 begins with the DNA derived from *Protocol 6*. The first step is a PCR that has been optimized for generating short (150–350 base) ^{32}P-labelled DNA probes from the reamplified DNA representing a candidate differential display band. The probe generated is then be used to screen a target cDNA library.

Protocol 9

Identification of full-length cDNAs clones represented by differential display bands

Reagents

- Reamplified DNA from differential display band (see *Protocol 6*)
- 25 μM T$_{11\ or\ 12}$VN oligonucleotide primer (see *Protocol 2*)
- 5 μM arbitrary 10-mer
- *Taq* polymerase and buffer (Promega)
- dNTP mix: 2.0 mM of dCTP, dGTP, dTTP; 0.2 mM of dATP
- [α-^{32}P]dATP, 3000 Ci/mmol (Du Pont)

Protocol 9 continued

Method

1. Combine the following components in a 0.5 ml microcentrifuge tube:
   - 2 µl of reamplified DNA
   - 2.5 µl of $T_{11 \text{ or } 12}VN$ primer
   - 2.5 µl of arbitrary 10-mer
   - 2.5 µl of 10 × Taq polymerase buffer
   - 2.5 µl of 25 mM $MgCl_2$
   - 2.5 µl of dNTP mix
   - 5 µl of [^{32}P]dATP (3000 Ci/mmol)
   - 1 µl (5 U) of Taq DNA polymerase
   - 4.5 µl of MQ H_2O

2. Perform PCR using cycling conditions described in *Protocol 5*.

3. Purify the probe using a G50 filtration column.

4. Hybridize the probe to a target cDNA library using standard library screening protocols. Use the probe at a final concentration of no less than 1×10^6 c.p.m./ml.

5. Plaque purify at least 12 individual cDNAs with each probe.

6. Sequence the 5′ and/or the 3′ region of each cDNA. Bona fide targets should be represented by at least 40% of the plaque-purified cDNAs.

7. Verify the expression pattern of each individual cDNA by either RNA gel blot analysis or RT-PCR.

The DNAs amplified from the indicated bands shown in *Figure 2* were used to identify their corresponding cDNAs using *Protocol 9*. In both cases at least half of the cDNAs identified in the library screen represented the same gene. The expression pattern of these genes was confirmed by RNA gel blot analysis and is shown in *Figure 4*.

5 Conclusion

Molecular approaches to distinguish mRNAs in comparative studies rely largely on differential screening or subtractive hybridization techniques (23, 24). Differential display provides a powerful and sensitive alternative and is now been widely used by many researchers. We describe in *Protocol 5* a modification which greatly reduces the signals generated by priming of the arbitrary 10-mer alone and thus selecting for poly(A)-containing cDNA fragments. By using this method, we have successfully isolated two novel seed-specific genes from *Arabidopsis thaliana* (19). In conclusion, differential display is sensitive and relatively easy, but requires efforts to filter out false positives. The frequency of false positive depends on many factors and may vary from experiment to experiment (25). A

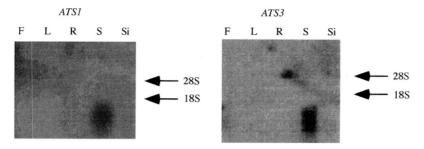

Figure 4 RNA gel blot analysis of *ATS1* and *ATS3*. RNA gel blots containing 10 μg of total RNA from flower (F), leaf (L), root (R), immature seed (S), and silique without seed (Si) were probed with *ATS1* and *ATS3*. The location of the 28S and 18S ribosomal RNAs are indicated. Figure and legend reproduced from *Figure 2A* in ref. 19 with kind permission from Kluwer Academic Publishers.

comprehensive knowledge of the intrinsic and extrinsic problems associated with differential display will greatly increase the success rate of this technique.

References

1. Liang, P. and Pardee, A. B. (1992). *Science*, **257**, 967.
2. Liang, P., Averboukh, L., and Pardee, A. B. (1993). *Nucleic Acids Res.*, **21**, 3269.
3. Liang, P., Zhu, W., Zhang, X., Guo, Z., O'Connell, R. P., Averboukh, L., *et al.* (1994). *Nucleic Acids Res.*, **22**, 5763.
4. Guimarães, M. J., Lee, F., Zlotnik, A., and McClanahan, T. (1995). *Nucleic Acids Res.*, **23**, 1832.
5. Vögeli-Lange, R., Bürckert, N., Boller, T., and Wiemken, A. (1996). *Nucleic Acids Res.*, **24**, 1385.
6. Zhang, H., Zhang, R., and Liang, P. (1996). *Nucleic Acids Res.*, **24**, 2454.
7. Wan, J. S., Sharp, S. J., Poirier, G. M.-C., Wagaman, P. C., Chambers, J., Pyati, J., *et al.* (1996). *Nature Biotechnol.*, **14**, 1685.
8. Benito, E. P., Prins, T., and van Kan, J. A. L. (1996). *Plant Mol. Biol.*, **32**, 947.
9. Martin-Laurent, F., van Tuinen, D., Dumas-Gaudot, E., Gianinazzi-Pearson, V., Gianinazzi, S., and Franken, P. (1997). *Mol. Gen. Genet.*, **256**, 37.
10. Crauwels, M., Winderickx, J., de Winde, J. H., and Thevelein, J. M. (1997). *Yeast*, **13**, 973.
11. Shen, W.-C. and Green, M. R. (1997). *Cell*, **90**, 615.
12. Gross, C. and Watson, K. (1998). *Yeast*, **14**, 431.
13. Sablowski, R. W. M. and Meyerowitz, E. M. (1998). *Cell*, **92**, 93.
14. Ausubel, S. F., Brent, R., Kingston, R. E., Moore, D. D., Seidman, J. G., Smith, J. A., and Struhl, K. (ed.) (1997). *Current protocols in molecular biology*. John Wiley, New York.
15. Chomczynski, P. and Sacchi, N. (1987). *Anal. Biochem.*, **162**, 156.
16. Callard, D. and Mazzolini, L. (1997). *Plant Physiol.*, **115**, 1385.
17. Brandstatter, I. and Kieber, J. J. (1998). *Plant Cell*, **10**, 1009.
18. Crouch, M. L., Tenbarge, K. M., Simon, A. E., and Ferl, R. (1983). *J. Mol. Appl. Genet.*, **2**, 273.
19. Nuccio, M. L. and Thomas, T. L. (1999). *Plant Mol. Biol.*, **39**, 1153.
20. Frugoli, J. A., Zhong, H. H., Nuccio, M. L., McCourt, P., McPeek, M. A., Thomas, T. L., *et al.* (1996). *Plant Physiol.*, **112**, 327.

21. Vielle-Calzada, J.-P., Nuccio, M. L., Budiman, M. A., Thomas, T. L., Burson, B. L., Hussey, M. A., *et al.* (1996). *Plant Mol. Biol.*, **32**, 1085.

22. Li, F., Barnathan, E. S., and Karikó, K. (1994). *Nucleic Acids Res.*, **22**, 1764.

23. Hedrick, S. M., Cohen, D. I., Nielsen, E. A., and Davis, M. M. (1984). *Nature*, **308**, 149.

24. Lee, S. W., Tomasetto, C., and Sager, R. (1991). *Proc. Natl. Acad. Sci. USA*, **88**, 2825.

25. Liang, P., Averboukh, L., and Pardee, A. B. (1994). In *Methods in molecular genetics* (ed. K. W. Adolph), Vol. 5, pp. 3–16. Academic Press.

Chapter 6

A modified approach for the efficient display of 3′ end restriction fragments of cDNAs

Y. V. B. K. SUBRAHMANYAM[†], SHIGERU YAMAGA, PETER E. NEWBURGER, and SHERMAN M. WEISSMAN

Department of Genetics, Boyer Centre for Molecular Medicine, Yale University School of Medicine, 295 Congress Avenue, New Haven, CT 06510, U.S.A.
[†]Gene Logic, Inc., 708 Quince Orchard Road, Gaithersburg, MD 20878, U.S.A.

1 Introduction

1.1 Gel display approaches to global analysis of gene expression

One of the remarkable developments in science in the last few years has been the progressive improvement in technology and paradigms for obtaining the sequence of very large amounts of genomic DNA. In parallel with this development, sequences identifying specific mRNAs have accumulated from random sequencing of segments of cDNA clones as well as from genomic sequences. The development of approaches for rapidly and economically screening the expression patterns of mRNAs in different tissues or cell types and as a function of different physiological or pharmacological perturbations promises to generate bodies of data of the same order of magnitude and biological significance as the sequences themselves.

Gene expression patterns can be used in a variety of ways to obtain detailed specific information about:

(a) Functions of individual genes.

(b) The commonality or divergence in expression controls between genes.

(c) The cell type specificity or generality of cellular responses to various stimuli.

(d) The regulation of gene expression during development and differentiation.

In addition, with appropriate systems, the global analysis of gene expression has the advantage that one will not miss important or even leading aspects of a class of regulatory phenomena simply by not using probes for an unsuspected gene product.

Analysis of gene expression in 'simple eukaryotes' (such as yeast, for instance) is relatively easier than in more complex ones. Simple organisms have fewer genes and it is easier to access cells in large numbers in the range of physio-logical states that evoke expression of most or all genes.

With the availability of the entire sequence of the yeast (*Saccharomyces cerevisiae*) genome (1, 2), it is clear that the yeast has less non-coding genomic DNA (so that recognition of template DNA is easier). There are fewer alternative splice forms of mRNA and fewer alternative promoters and gene families that are simpler with less partially redundant sequence. Thus displaying the 6000 or so yeast genes (3), either as synthetic oligonucleotides (4) or as immobilized DNA fragments (5) in arrays that can be rapidly analysed by hybridization, is technically speaking a more tractable proposition. In contrast, the global analysis of gene expression in complex systems such as mammalian cells for instance is technically more challenging. Presently it is estimated that over 80% of all expressed sequences are represented in the database of expressed sequenced tags (ESTs) (6). However a number of products are represented by one or two ESTs. A number of ESTs contain repetitive sequences that may or may not be derived from the 3′ end of cytoplasmic transcripts. There may be cases in which non-overlapping ESTs (or clones) are derived form alternative 3′ ends of a message (7, 8). The preparation of arrays (9–11) with such a diversified group of ESTs on a routine, stable, and affordable basis has yet to be fully demonstrated. For these and other related reasons as of now it appears that to get a full picture of the pattern of gene expression in a mammalian cell it is necessary to use methods which are independent of the prior knowledge of all genes that is required for arrays.

In addition to array analysis, two broad types of approaches have been used for the global analysis of gene expression patterns. One approach consists of obtaining sequence from a sufficient number of randomly chosen clones from a cDNA library. This approach has been considerably expedited by the SAGE (serial analysis of gene expression) method (12) of analysis in which short nucleotide sequences are obtained and concatenated prior to sequencing, so that one sequence run covers a number of cDNA molecules. The other approach has consisted of parsing cDNA fragments into subsets such that each subset is sufficiently simple that the members of a subset can be separately seen and analysed following fractionation on conventional sequencing gels (13, 14). We have successfully used a similar gel-based display method in which the cDNA fragments to be displayed are derived by restriction enzyme digestion of the cDNA (15). In this way the fragment sizes are predictable from the sequence of the cDNA. Also PCR amplification can be performed under stringent conditions so that essentially all the amplified products have the 'correct' sequence.

It is desirable to use a method that could demonstrate levels of unknown RNA species as well as those already represented in the databases, and that could conveniently be applied to multiple RNA samples. We have developed such a technique by modification of the previously described '3′ end cDNA restriction fragment display analysis' method (15).

This modified method of differential display has proven to be very easy and useful for uncovering changes of mRNA as well as detecting mRNAs not so far documented in the public databases. Changes of about two- to threefold in the relative levels of individual amplification products could commonly be detected and mRNAs whose relative abundance ranged over three to four orders of magnitude could be measured. Sequences from the cDNA bands could be readily obtained in most cases, and were useful not only in identifying unique sequence mRNAs, but also in distinguishing RNA species containing common repetitive elements. This chapter describes the detailed protocol that we used in most of our studies.

The general strategy of the modified approach (16) for the display of 3' end restriction fragments of cDNAs is outlined in *Figure 1*. Total RNA isolated from a given experimental sample is primed for first strand cDNA synthesis (17), using an oligo(dT)$_{18}$ primer that carries a T7-Sal heel sequence on the 5' end. The primer contains equimolar amounts of oligonucleotides that carry A or C or G as N1 at the 3' end (T7-Sal-oligo(dT)$_{18}$V; where V is A, C, or G) (see also Section 2.1.1). The second strand synthesis is carried out by the Gubler and Hoffman method (18). The cDNA prepared is digested with a six nucleotide recognizing restriction enzyme and the products are ligated to the Y-shaped 'Fly-Adapter' (see Section 2.3). The ligated cDNA is subjected to PCR amplification using ^{32}P-labelled FA-1 primer as the 5' primer and the N1N2 anchored oligo(dT)$_{18}$ primer that has a heel different from T7-Sal (see Section 2.4.2 and Protocol 8), as the 3' primer. As there are 12 different N1N2 anchored primers possible (where N1 is a non-T residue), for every ligated cDNA 12 individual PCR reactions are carried out. Only the 3' end fragments of the cDNA can participate in the PCR amplification, although the internal fragments and the 5' end fragments of the cDNA generated by the restriction enzyme do get ligated to the Fly-Adapter. This is because only the 3' end fragments carry the oligo(dT)$_{18}$V as a part of the cDNA and during the first cycle of PCR the 5' primer cannot anneal to the template and hence doesn't participate in the PCR amplification. On the other hand the internal and 5' end cDNA fragments do not have the oligo(dT)$_{18}$ region and hence theoretically cannot result in any amplifiable template. The PCR amplified products are subsequently analysed by electrophoresis on sequencing gels and the bands are visualized by autoradiography (see *Figure 3*).

2 Methods

2.1 cDNA synthesis

One of the important issues to be considered prior to starting the cDNA synthesis reaction is the quality of RNA. Isolation of high quality RNA is the key for the success in getting good cDNA and subsequent reliable band pattern in the display analysis. Several protocols and kits are available to isolate good quality intact RNA with ease and reproducibility. The choice of RNA isolation method depends on the source of the starting material. Each laboratory uses a different method that works well in their hands. We use TRIzol reagent (Gibco BRL) for isolation of RNA from several cell lines. However for neutrophil RNA

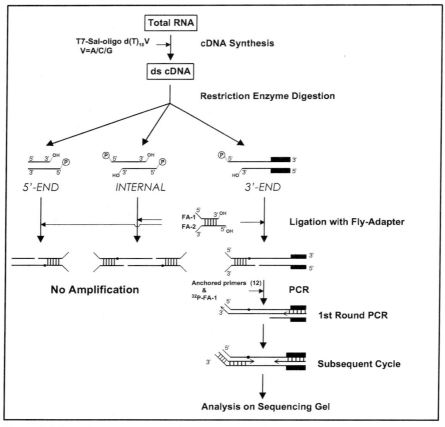

Figure 1 Schematic representation of the 'display of 3' end restriction fragments of cDNAs' method.

we used the method described by Strohman *et al.* (19) with some modifications as described (16). It is important to ensure the quality and integrity of the total RNA by analysing an aliquot of the preparation on a formaldehyde-agarose gel (20) and assessing relative intensities of 28S and 18S ribosomal RNA bands.

The RNA preparations sometimes may contain varying amounts of genomic DNA. This may lead to spurious bands on the display gels due to the non-specific priming from the poly(A) stretches on the DNA. Therefore it is required to ensure that the RNA preparation is free of this genomic DNA. This can be achieved by incubating the RNA preparation with DNase I that is free of RNase contamination followed by phenol extraction and ethanol precipitation.

According to the original procedure described by Prashar and Weissman (15), the RNA was primed for first strand cDNA synthesis with an anchored oligo(dT) primer that has a heel sequence. The anchor used was a dinucleotide (N1N2) (13, 14) where the N1 is not a T. Therefore 12 individual cDNA preparations are to be made from every RNA sample by priming separately with each of the 12 N1N2 primers. This requires a substantial amount of RNA (120 µg). Sometimes it is not possible to obtain sufficient amounts of RNA from every experimental condition or from the clinical sample we would want to analyse. We attempted to overcome this difficulty by priming the RNA with oligo(dT)$_{18}$V (where V is A, C, or G), which means the cDNA is made with a mixture of A/C/G anchors. Therefore for every RNA sample we need to make only one cDNA preparation. The first strand cDNA synthesis reaction was carried out at a lower temperature (45°C) in order to provide the required specificity for the N1 anchored primer. The cDNA made by this approach is digested with a restriction enzyme, ligated to the Fly-Adapter as described (see Section 2.3), and subjected to PCR amplification. Each of the N1N2 anchored primers was used at the PCR stage to set up 12 individual PCR reactions. Approximately 6 μg of total RNA was used to make each cDNA. The amount of cDNA made from 6 μg of RNA is enough to analyse the display patterns with about ten restriction enzymes using all the anchored primers. Apart from being able to use small amounts of RNA for the analysis, this approach very conveniently avoids the need to make a large number of cDNA preparations and associated tube-to-tube variations that can occur during pre- and post-cDNA synthesis steps.

A method has also been developed in our laboratory that uses much smaller amounts of RNA (much less than a microgram) for cDNA synthesis and subsequent display analysis (21). This method employs pre-amplification of the cDNA by ligating adapters on either end followed by a modified PCR step. A detailed protocol has been described elsewhere (21).

The first strand synthesis can also be performed at higher temperatures with superscript II reverse transcriptase, by using thermostabilizing compounds such as D(+)trehalose (22, 23) or by using thermostable reverse transcriptase (Gibco BRL) enzyme in order to increase the specificity of the annealing with the anchored oligo(dT) primer. This may also help in avoiding non-specific priming at the internal poly(A) stretches.

2.1.1 First strand synthesis

About 5–6 μg of total RNA is used for every cDNA synthesis reaction (17). This results in enough cDNA to generate display patterns with ten restriction enzymes.

2.1.2 Second strand synthesis

At the end of the first strand synthesis chill the tubes on ice and centrifuge briefly to collect all the contents to the bottom of the tube. It is important to ensure that the tubes do not stay on ice for a prolonged time at the end of first strand synthesis reaction.

It is desirable to carry out some parallel cDNA synthesis reactions under identical conditions that can be used to follow up the first and second strand synthesis (separately) steps by using the incorporation of [α-^{32}P]dCTP (3000 Ci/mmole) as described in detail (17). By this approach we can calculate the efficiency of cDNA synthesis. The ^{32}P-labelled first and second strand cDNA samples can be subjected to alkaline agarose gel electrophoresis (20) to assess the quality and size of the cDNA strands made in the reactions.

Protocol 1

First strand cDNA synthesis

Reagents

- Diethyl pyrocarbonate (DEPC) treated water
- Superscript II reverse transcriptase (200 U/µl) along with 5 × first strand buffer (250 mM Tris–HCl pH 8.3, 375 mM KCl, 15 mM MgCl$_2$) and 0.1 M dithiothreitol (DTT) (Gibco BRL)
- RNase inhibitor (rRNasin, 40 U/µl, Promega Corporation)
- 100 mM deoxyribonucleoside triphosphates (dNTP) (New England Biolabs)
- T7-Sal-oligo(dT)$_{18}$V: 5′-ACG TAA TAC GAC TCA CTA TAG GGC GAA TTG GGT CGA C TTT TTT TTT TTT TTT TTT V-3′ (V = A/C/G)

Method

1. Take 6 µg of total RNA in 10 µl volume of DEPC treated water in a 0.5 ml micro-centrifuge tube (Low Adhesion Microcentrifuge tubes, USA Scientific Plastics) on ice. Add 1 µl (200 ng) of T7-Sal-oligo(dT)$_{18}$V as the primer for the reaction. Mix the contents on ice.

   - 6 µg RNA in water   10 µl
   - 200 ng T7-Sal-oligo(dT)$_{18}$V   1 µl
   - Final volume   11 µl

2. Incubate for 5 min at 65 °C, and chill it for 5 min on ice. Repeat this denaturation and annealing once again to make sure that the RNA secondary structures are denatured. Keep the tubes on ice.

3. Add the following components to the tube (on ice) of the first strand cDNA synthesis reaction (final volume = 19 µl) as follows:[a]

   - 5 × first strand buffer   4 µl
   - 0.1 M DTT   2 µl
   - 10 mM dNTPs   1 µl
   - RNase inhibitor (40 U/µl)   1 µl
   - Volume added   8 µl

4. Pre-warm the contents to 45 °C and initiate the cDNA synthesis by adding 1 µl (200 U) of Superscript II reverse transcriptase.

5. Continue incubation (at this stage the final reaction volume is 20 µl) at 45 °C for 1 h.[b]

[a] If there are several RNA samples make a reaction master mix with the above four components and dispense 8 µl per tube to maintain consistency.

[b] Evaporation during incubation is a significant problem and it can be dealt with in several ways. This step can either be performed in an air incubator at 45 °C or a mineral oil overlay can be used if a water-bath or PCR machine is used for incubation.

Protocol 2

Second strand cDNA synthesis

Reagents

- 5 × second strand buffer: 100 mM Tris–HCl pH 6.9, 23 mM MgCl$_2$, 450 mM KCl, 0.75 mM β-NAD$^+$, 50 mM (NH$_4$)$_2$SO$_4$ (Gibco BRL, 10812-014)
- DNA polymerase I (10 U/μl), E. coli DNA ligase (10 U/μl), and ribonuclease H (3 U/μl) (Gibco BRL)

Method

1. Set up the second strand reaction on ice into the same tube as described below.[a]

   - First strand reaction   20 μl
   - Water   91 μl
   - 5 × second strand buffer   30 μl
   - 10 mM dNTPs   3 μl
   - E. coli DNA polymerase (10 U/μl) 4 μl
   - E. coli DNA ligase (10 U/μl)   1 μl
   - RNase H (3 U/μl)   1 μl
   - Final volume   150 μl

2. Incubate the tubes at 16°C for 2 h.

[a] For multiple samples a master mix can be made on ice in a 1.5 ml microcentrifuge tube (Low Adhesion Microcentrifuge tubes). The enzymes are added to the master mix and mixed gently just prior to use. Aliquot 130 μl of the master mix to each tube containing the first strand reaction sample (20 μl) to give a final volume of 150 μl. Mix gently and incubate at 16°C for 2 h.

Protocol 3

cDNA clean-up

Reagents

- TE buffer: 10 mM Tris–HCl pH 8.0, 1 mM EDTA
- Tris–HCl buffer saturated phenol (Gibco BRL)
- 20 mg/ml glycogen (Boehringer Mannheim)

Method

1. At the end of incubation, add 10 μl of 0.5 M EDTA pH 8.0 to stop the reaction.

2. Extract the cDNA once with phenol:chloroform (1:1, v/v) and once with chloroform.[a]

Protocol 3 continued

3. Add 0.5 vol. of 7.5 M ammonium acetate to the aqueous phase and precipitate[b] the cDNA with 2.5 vol. of ethanol (pre-chilled to −20°C). At this stage the sample can be left overnight at −20°C.

4. Collect the cDNA precipitate by centrifugation in an Eppendorf centrifuge at top speed for 15 min.

5. Carefully remove the ethanol supernatant (without disturbing the pellet). Wash the pellet with 70% (v/v) ethanol and centrifuge again for 15 min.

6. Remove the supernatant and air dry the pellet at room temperature.

7. Dissolve the cDNA in 20 μl of water or TE buffer.

[a] Pre-saturate the chloroform with DEPC treated water.

[b] Prior to precipitation of the cDNA 1 μl (20 μg) of glycogen may be added as a carrier.

2.2 Restriction enzyme digestion

The cDNA prepared as described above can be digested separately with a variety of six nucleotide recognizing restriction enzymes based on their frequency of cutting on a given cDNA (such as murine or human). The enzymes that recognize four nucleotide sequence digest the cDNA at much higher frequency. As a result the number of 3′ end fragments generated are so many that they can not be resolved into discrete bands on the sequencing gel.

The procedure described below is for digesting the cDNA with *Xba*I enzyme, however we can use any restriction enzyme with its corresponding 10 × buffer to set up the digestion in the same way.

Protocol 4

Restriction enzyme digestion

Reagents

• 10 × buffer (10 × buffer No. 2): 100 mM Tris–HCl pH 7.9, 100 mM MgCl$_2$, 500 mM NaCl, 10 mM dithiothreitol (New England Biolabs)

• Restriction enzyme *Xba*I (New England Biolabs)

Method

1. Prepare a reaction master mix on ice as follows:[a]

    • Water   33 μl
    • 10 × enzyme (*Xba*I) buffer   5 μl mix well and add
    • Restriction enzyme (*Xba*I)   2 μl (20 U)
    • Final volume   40 μl

2. Set up the digestion of the cDNA by adding 8 μl of the reaction master mix to 2 μl of cDNA and incubate the sample at 37 °C for 2 h.

3. At the end of incubation heat inactivate the enzyme at 65 °C for 20 min, and keep the tubes on ice.

a Always prepare the reaction master mix freshly just prior to use and discard the remaining.

2.3 Ligation of digested cDNA with Fly-Adapter

These Y-shaped adapters are prepared by annealing the adapter oligonucleotides FA-1 and FA-2 (see Protocol 5). This Fly-Adapter has a region where both FA-1 and FA-2 are annealed together, and a region where they do not anneal, as these oligonucleotides are non-complementary in this region. This results in generating an Y-shape to the adapter (see *Figure 2*).

The annealed Fly-Adapter has a four nucleotide overhang, which is complementary to the end generated by the restriction enzyme. Fly-Adapters compatible with the other restriction enzymes of interest can be designed accordingly by altering the four variable overhang bases of the FA-2. For the enzymes that result in a 3' overhang such as *Pst*I, the corresponding four base overhang is present on FA-1 instead of on FA-2. In either case we could use ^{32}P-labelled FA-1 for PCR amplification.

In the original procedure (15), the FA-2 adapter was phosphorylated prior to annealing of the adapter oligonucleotides. This was to ensure that both FA-1 and FA-2 are ligated to the 5' overhang of the 3' end restriction fragments of the cDNA. This step is not required and in the present procedure at the end of ligation only the FA-1 gets ligated to the cDNA (see *Figure 2*). This works equally well.

There are some restriction enzymes that do not get heat inactivated sub-

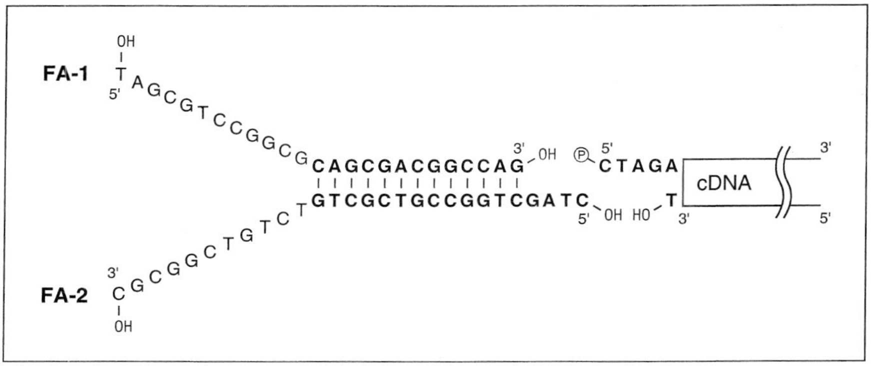

Figure 2 Diagrammatic representation of the annealed adapter with 5'-CTAG overhang and the 3' end restriction fragment of the cDNA with the compatible overhang (5'-CTAG, generated by the restriction enzyme *Xba*I).

sequent to digestion therefore it is desirable to design the overhang on FA-2 such that once the Fly-Adapter is ligated to the cDNA it does not regenerate the enzyme site.

Protocol 5

Annealing of the Fly-Adapter oligonucleotides FA-1 and FA-2

Reagents

- 10 × annealing buffer: 100 mM Tris–HCl pH 8.0, 10 mM EDTA pH 8.0, 1 M NaCl
- FA-1 (Fly-Adapter 1): 5'-TAG CGT CCG GCG CAG CGA CGG CCA G-3'
- FA-2 (Fly-Adapter 2):[a] 5'-**CTAG** CTG GCC GTC GCT GTC TGT CGG CGC-3'

Method

1. Take 5 μg each of adapter oligonucleotides FA-1 and FA-2 into a 1.5 ml micro-centrifuge tube (Low Adhesion Microcentrifuge tubes).

2. Add 10 μl of 10 × annealing buffer and adjust the final volume with water to 100 μl. At this stage the concentration of the Fly-Adapter is 100 ng/μl.

3. Heat the sample in a boiling water-bath for 5 min. Turn off the burner, allow the bath to reach room temperature, and then place at 4°C.

4. Collect the contents in the bottom of the tube by a brief centrifugation.

5. Store the annealed Fly-Adapter at −20°C.

[a] The overhang nucleotides on FA-2 were designed for the enzyme *Xba*I. By altering the overhang bases of the FA-2 we can design the Fly-Adapters suitable for other restriction enzymes of interest.

Protocol 6

Ligation of cDNA with Fly-Adapter

Reagents

- 10 × ligase buffer: 500 mM Tris–HCl pH 7.8, 100 mM MgCl$_2$, 100 mM DDT, 10 mM ATP, 250 μg/ml bovine serum albumin (New England Biolabs)
- T4 DNA ligase (400 U/μl) (New England Biolabs)

Method

1. Prepare the ligase master mix as follows:

   (a) Take a 0.5 ml microcentrifuge tube on ice, add 88 μl of DEPC treated water, and 10 μl of 10 × T4 DNA ligase buffer. Mix the contents in the tube gently, add 2 μl of T4 DNA ligase (400 U/μl), mix again gently. This is diluted ligase.[a]

    (b) In another tube on ice take 20 μl of water, 10 μl of 10 × T4 DNA ligase buffer, mix well, and add 10 μl of diluted ligase (final volume at this stage is 40 μl). This is referred to as ligase master mix.

2. Set up the ligase reaction as follows. Take 2 μl of restriction enzyme (XbaI) digested cDNA in a 0.5 ml microcentrifuge tube on ice and add 1 μl of Fly-Adapter with suitable overhang (5′-CTAG overhang for XbaI). Mix gently and add 2 μl of ligase master mix.[a]

3. Mix the contents and incubate at 16 °C overnight.

[a] Always prepare the diluted ligase and the ligase master mix freshly just prior to use and discard the remaining.

Commercially available T4 DNA ligase enzyme is very concentrated. As we are dealing with very small amounts of the digested cDNA and adapter the ligation procedure includes a step to dilute the enzyme to a range that is not in excess. We generally use 4 U of ligase in a 5 μl final volume of the reaction.

2.4 PCR amplification

For every adapter ligated cDNA sample, 12 PCR reactions are set up using N1N2 anchored oligo primer as the 3′ primer and ^{32}P-labelled FA-1 as the 5′ primer. We use a two stage PCR protocol for the PCR amplification of the Fly-Adapter ligated cDNA. The first stage of PCR has five cycles of annealing at 55 °C. This low annealing temperature helps the anchored N1N2 primers to anneal to the template efficiently. Following this in the second stage we increase the annealing temperature to 60 °C to help maintain the specificity. We have found it desirable to use the T7-Sal-oligo(dT)$_{18}$V primer for cDNA synthesis. We used AmpliTaq Gold for all our display analysis. This enzyme resulted in clean patterns than the regular AmpliTaq enzyme. Platinum Taq polymerase from Gibco BRL was also tested. We found that this enzyme requires lesser amounts of template and gives a better amplification of the high molecular weight cDNA bands. Hence this enzyme is also a good choice for the display analysis with some optimization. Both of the enzymes are convenient in setting up hot start PCR reactions.

One of the critical aspects to consider prior to setting up the PCR reaction is optimization of the amount of template cDNA (Fly-Adapter ligated) to be used. Too much of template results in smeary patterns while too little template results in faint and/or missing cDNA bands. Therefore it is desirable to try different dilutions and select the optimal dilution and quantity of the template for the PCR amplification. Generally when we start with about 6 μg of total RNA (we consistently get about 30% efficiency of cDNA synthesis), 2 μl of 1:20 diluted template works well for a 20 μl PCR reaction.

Even under identical PCR conditions we observed that there is some variation in the incorporation of radioactivity with respect to the anchored primers used. While the PCR conditions described above work well for all the anchored primers, it is possible to employ higher annealing temperature or reduce the number of cycles with some of the anchored primers (as in case of GG, GC, CG, and CC). The annealing temperature could be lowered or the number of cycles may be increased for GT, AT, CT, and GA anchored primers.

Protocol 7

^{32}P-labelling of the primer FA-1

Equipment and reagents

- Quick Spin™ (Sephadex G25, fine) columns (Boehringer Mannheim)
- 10 × kinase buffer: 700 mM Tris–HCl pH 7.6, 100 mM MgCl$_2$, 50 mM DTT (New England Biolabs)
- [γ-^{32}P]ATP (3000 Ci/mmole) (Amersham Corporation)
- T4 polynucleotide kinase (T4PNK, 10 U/μl) (New England Biolabs)

Method

1. Take a 0.5 ml microcentrifuge tube on ice, add 2 μl (1 μg) of FA-1 primer, and set up the reaction as follows:

   - 1 μg FA-1 primer   2 μl
   - 10 × T4 polynucleotide kinase buffer   2 μl
   - 150 μCi [γ-^{32}P]ATP (3000 Ci/mmole)   15 μl
   - T4 polynucleotide kinase (10 U/μl)   1 μl
   - Final volume   20 μl

2. Incubate at 37°C for 1 h. At the end of the incubation add 20 μl of water to the tube.

3. Heat inactivate the kinase enzyme at 65°C for 15 min, chill the tube on ice, and centrifuge briefly.

4. Purify the ^{32}P-labelled primer FA-1 from unincorporated [γ-^{32}P]ATP using a Quick Spin (Sephadex G25, fine) column following the instructions of the supplier.

5. Add 1 μg of unlabelled FA-1 to the purified ^{32}P-labelled primer (FA-1, approximate volume at this stage is 40 μl) and adjust the final volume to 80 μl.

6. At this point the concentration of this primer is 25 ng/μl and is ready for use in PCR.

The PCR reactions (final volume 20 μl) are carried out in 0.5 ml thin-wall PCR tubes as described below. For every Fly-Adapter ligated cDNA, there will be a set of 12 tubes one for each of the 12 anchored primers.

At the end of the PCR, the products are ready for analysis on a 6% poly-acrylamide–7 M urea sequencing gel. Although for every PCR reaction with each anchored primer we use same amount of the template and ^{32}P-labelled primer FA-1, there is some variation among these anchored primers in terms of the amount of radioactivity incorporated. For evaluating the differences associated with the expression of individual cDNAs under different experimental conditions and for accurate quantitation using a PhosphorImager, it is desirable to load equal amount of radioactivity for each sample. This can be achieved by purifying each of the PCR amplified cDNA. Purification of the PCR products also reduces the background.

Protocol 8

PCR reaction

Reagents

- AmpliTaq Gold (5 U/μl) (Perkin Elmer)
- GeneAmp 10 $\times$ PCR buffer: 100 mM Tris–HCl pH 8.3, 500 mM KCl, 15 mM MgCl$_2$, 0.01% (w/v) gelatin (Perkin Elmer)
- Nujol mineral oil (Perkin Elmer)
- Anchored oligo(dT)$_{18}$ N1N2 primers and corresponding heel sequences as listed below:

RP 8.0 AA	5'-**TGA AGC CGA GAC GTC GGT CG** TTT TTT TTT TTT TTT TTT **AA**-3'
RP 8.0 AC	5'-**TGA AGC CGA GAC GTC GGT CG** TTT TTT TTT TTT TTT TTT **AC**-3'
RP 8.0 AG	5'-**TGA AGC CGA GAC GTC GGT CG** TTT TTT TTT TTT TTT TTT **AG**-3'
RP 8.0 AT	5'-**TGA AGC CGA GAC GTC GGT CG** TTT TTT TTT TTT TTT TTT **AT**-3'
RP 5.0 CA	5'-**CTC TCA AGG ATC TTA CCG CT** TTT TTT TTT TTT TTT TTT **CA**-3'
RP 5.0 CC	5'-**CTC TCA AGG ATC TTA CCG CT** TTT TTT TTT TTT TTT TTT **CC**-3'
RP 5.0 CG	5'-**CTC TCA AGG ATC TTA CCG CT** TTT TTT TTT TTT TTT TTT **CG**-3'
RP 5.0 CT	5'-**CTC TCA AGG ATC TTA CCG CT** TTT TTT TTT TTT TTT TTT **CT**-3'
RP 6.0 GA	5'-**TAA TAC CGC GCC ACA TAG CA** TTT TTT TTT TTT TTT TTT **GA**-3'
RP 6.0 GC	5'-**TAA TAC CGC GCC ACA TAG CA** TTT TTT TTT TTT TTT TTT **GC**-3'
RP 6.0 GG	5'-**TAA TAC CGC GCC ACA TAG CA** TTT TTT TTT TTT TTT TTT **GG**-3'
RP 6.0 GT	5'-**TAA TAC CGC GCC ACA TAG CA** TTT TTT TTT TTT TTT TTT **GT**-3'

Method

1. Take a 0.5 ml thin-wall PCR tube on ice, add 2 μl of a 2 μM stock solution of 3' anchored primer (N1N2 anchored oligo(dT)$_{18}$ with heel, see reagents), and mix with 18 μl of the reaction master mix.[a] Overlay the tube with mineral oil.

2. The reaction master mix will have the following components per tube:
   - GeneAmp 10 $\times$ PCR buffer (with MgCl$_2$)   2 μl
   - 2 mM dNTPs   2 μl
   - ^{32}P-labelled primer FA-1   2 μl
   - Water   9 μl

   At this stage mix everything gently then add:
   - AmpliTaq Gold (5 U/μl)   0.7 U/reaction
   - Fly-Adapter ligated cDNA template   2 μl of 1:20 diluted cDNA in water
   - Adjust the final volume to   18 μl

3. PCR conditions.

   (a) Stage I: one cycle of 94°C for 12 min.

Protocol 8 continued

(b) Stage II: five cycles of 94 °C for 30 sec; 55 °C for 2 min; 72 °C for 1 min.

(c) Stage III: 25 cycles of 94 °C for 30 sec; 60 °C for 2 min; 72 °C for 1 min.

4. At the end of PCR the samples can be stored at −20 °C, prior to gel analysis.

[a] It is highly recommended to make a reaction master mix based on the total number of reactions to maintain consistency and to avoid tube-to-tube variation. While preparing the master mix, add all the required components except the enzyme and mix gently. Add the enzyme last, gently mix again, give a brief centrifugation, and dispense 18 μl each into individual tubes.

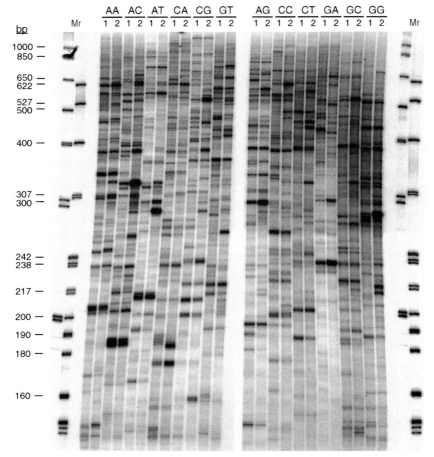

Figure 3 Display analysis of differential mRNA expression in neutrophils. Total RNA isolated from neutrophils was primed with oligo(dT)$_{18}$V to synthesize cDNA. This cDNA was digested with *Xba*I enzyme, ligated to Fly-Adapter, and subjected to differential display using the anchored primers as described in the text. Panels A and B represent gel runs covering all the anchored primers. Lane 1, untreated (control) neutrophils; lane 2, neutrophils incubated with opsonized *E. coli*. The N1N2 anchors used are indicated on the top of the lanes. Two different molecular weight markers (1 kb Plus and pBR322 digested with *Msp*I) were used as size standards.

This purification is time-consuming and cumbersome particularly when there are a large number of samples, and hence can be treated as an optional step. We generally analyse the PCR products directly without purification (see *Figure 3*).

Protocol 9

Purification of the PCR products

Reagents

- PCR Purification Kit (Qiagen), used according to the instructions provided with the kit

Method

1. At the end of PCR, to purify the sample use a PCR Purification Kit (Qiagen) following the instructions provided by the supplier.

2. Elute the products from the column into a volume of 30 μl with the elution buffer supplied with the kit.

3. Use 1 μl of the sample to determine the counts in a scintillation counter.

4. Load equal amounts of the radioactivity (in terms of Cerenkov counts) on the gel.

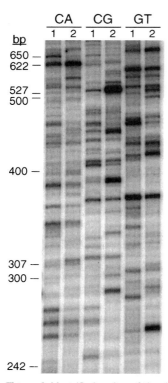

Figure 4 Magnified region of the display gel. A selected region for CA, CG, GT anchors from *Figure 3* ranging from 622 bp–242 bp is enlarged to show the differentially regulated bands in detail. Lanes 1 and 2 are as described in *Figure 3*.

2.5 Sequencing gel analysis of the PCR products

The PCR amplified products are analysed on a 6% polyacrylamide–7 M urea sequencing gel. It is important to load ^{32}P-labelled molecular weight markers along with the samples. We routinely use ^{32}P-labelled 1 kb Plus ladder (Gibco BRL) or *Msp*I digested pBR322 (New England Biolabs) as the markers. Alternatively, sequencing ladders can also be used to size the PCR products on the gel. The marker DNA can be conveniently labelled using [γ-^{32}P]ATP by an exchange reaction with T4 polynucleotide kinase.

Protocol 10

^{32}P-labelling of molecular weight marker DNA

Reagents

- T4 polynucleotide kinase (Gibco BRL)
- 5 × exchange reaction buffer: 250 mM imidazole–HCl buffer pH 6.4, 60 mM MgCl$_2$, 5 mM 2-mercaptoethanol, 350 μM ADP (Gibco BRL)
- 1 kb Plus ladder (Gibco BRL)
- pBR322/*Msp*I digest (New England Biolabs)
- [γ-^{32}P]ATP: 3000 Ci/mmole (Amersham Corporation)
- Quick Spin™ (Sephadex G25, fine) columns (Boehringer Mannheim)

Method

1. Take a 0.5 ml microcentrifuge tube on ice, add 12 μl water, 5 μl of 5 × exchange reaction buffer, 2 μl (2 μg) of 1 kb Plus marker DNA, and 5 μl (50 μCi) of [γ-^{32}P]ATP.
2. Initiate the reaction by adding 1 μl (10 U) of T4 polynucleotide kinase enzyme followed by incubation at 37 °C for 30 min.
3. Heat inactivate the kinase enzyme at 65 °C for 15 min.
4. Remove the unincorporated [γ-^{32}P]ATP using a Quick Spin (Sephadex G25 fine) column.
5. Take 1 μl of the ^{32}P-labelled marker, add 5 μl of stop solution, and 4 μl of water.
6. Denature the DNA at 95 °C for 3 min (at least) and chill it on ice. Load 2 μl of this sample as a marker on the gel.

Protocol 11

Preparation of the sequencing gel and electrophoresis

Reagents

- Stop solution (once prepared this can be used for a month): 95% (v/v) formamide, 20 mM EDTA pH 8.0, 0.05% (w/v) bromophenol blue, 0.05% (w/v) xylene cyanol FF
- 1 × TBE buffer (20): 90 mM Tris–borate pH 8.3, 2 mM EDTA
- Sequencing gel mix-6 (Gibco BRL)
- Sigmacote (Sigma)
- Ammonium persulfate and *N,N,N',N'*-tetramethylethelenediamine (TEMED) (Bio-Rad Laboratories)
- Deionized formamide (Gibco BRL)

Method

1. Rinse the sequencing gel plates (Gibco BRL, S2001 sequencing system) with ethanol once and let the plates dry for a few minutes. Apply Sigmacote[a] (about 0.5 ml) on the smaller of the two gel plates, spread it uniformly on the plate with the help of Kimwipes, and let dry for a few minutes. This will siliconize the surface of the glass plate.

2. Rinse the small plate with ethanol once again following siliconization, and allow the plates to dry for a few minutes.

3. Using 0.4 mm spacers and a gel casting boot assemble the glass plates as described in the instruction manual provided by the supplier.

4. Cast the gel using gel mix-6 acrylamide gel solution. The comb has 6.8 mm wide teeth and 2 mm space between the teeth (36-wells, this can be custom-made from Gibco BRL).

5. Once the gel is polymerized remove the comb carefully and assemble the gel in the electrophoresis apparatus that has a cooling base attached. The electrophoresis buffer is $1 \times$ TBE. Load 2 µl of stop solution each in a couple of wells and do a pre-electrophoresis[b] for about 1–2 h at 2200 V constant voltage.

6. Take 2 µl of the PCR product, mix with an equal volume of stop solution, and prepare the samples for loading by heat denaturation[c] at 95 °C for 3 min, followed by chilling them quickly on ice.

7. At the end of the pre-electrophoresis disconnect the power supply to the gel unit and wash the wells using a syringe to remove the diffused urea from the wells. Load the samples with flat-tipped sample loading tips designed for the sequencing gel. Continue the electrophoresis at 2200 V constant voltage.

8. Continue the electrophoresis till the xylene cyanol dye reaches the bottom of the gel or has just run out of the gel.

9. Remove the gel at the end, separate the glass plates, and transfer the gel on to a pre-cut sheet of filter paper (Whatman 3MM) by placing the filter paper on the top of the gel surface and gently rubbing with Kimwipes (extra long) uniformly.

10. Carefully lift the filter paper along with the gel and place the gel side up.

11. Cover the gel surface with extra wide Saran Wrap (this may be purchased from wholesale warehouse clubs such as Sam's club).

12. Dry the gel in a gel drier. It is important to have a second layer of pre-cut sheet of filter paper (Whatman 3MM) under the gel to protect the gel from picking up any contaminants from the gel drier.

[a] Wear gloves while applying Sigmacote and it is desirable to do this siliconization of the glass plate in a fume hood.

[b] It is important to make sure that the gel temperature is in the range of 45 °C–50 °C. This should allow the samples to remain denatured during the run.

[c] This can be done by taking the samples into 0.5 ml PCR tubes and heating the tubes in a PCR machine.

Protocol 12

Autoradiography

Equipment

- BioMax MR X-ray film (Kodak)
- Radtape (Diversified Biotech)

Method

1. Once the gel is dried and cooled to room temperature carefully remove the gel from the drier.

2. Trim the edges of the filter paper sheet and seal the edges of the gel along with the Saran Wrap using a Scotch tape. This helps in keeping the wrap intact on the gel surface.

3. Cut a few pieces of Radtape. Write some details about the gel along with some arrow marks on the tape and stick them at several places along the edges of the gel (this will help in aligning the X-ray film on the dried gel for recovering the cDNA bands of interest for further analysis).

4. Expose the gel to a BioMax-MR X-ray film overnight at room temperature.

5. Store the dried gels carefully so that they can subsequently be used for recovering the cDNA bands.

6. It is also important to make sure that these gels do not shrink or accumulate any contaminants and/or dirt during storage[a] that could potentially cause problems in subsequent PCR amplifications.

[a] Zip-seal polyethylene bags (13 × 18 inches, 4 mil. thickness, PGC Scientific, 2-3321-18) can be used for this purpose.

2.6 Recovery of differentially regulated cDNA bands from the dried gels for sequencing

Specific bands of interest from the display gel based on the autoradiogram can be excised (15) from the gel and reamplified for further analysis. It is desirable to use the gel within a few days for recovering the bands of interest as prolonged storage of the gels might lead to shrinkage of the gel and thereby create difficulty in aligning the gel with the X-ray film.

The cDNA band eluted from the gel may still have some trace amounts of other cDNAs as contaminants which may get co-amplified during PCR. It is important to remove these contaminants by another round of PCR amplification with the ^{32}P-labelled FA-1 and the anchored primer. At this stage we can also use the corresponding heel primer in the place of the anchored primer. The products are resolved on a sequencing gel.

The PCR products are analysed on a sequencing gel and the specific cDNA bands are recovered again. At the end of these steps the cDNA bands are pure enough for further analysis.

Protocol 13

Recovery of the cDNA bands

Equipment and reagents

- Sharp razor blade
- 1.5 ml microcentrifuge tubes
- TE buffer
- Glycogen
- NaOAc pH 6.0
- Ethanol

Method

1. Identify the bands of interest and mark them with a fine-point marker.

2. Mount the gel on a clean soft wooden board covered with a couple of sheets of filter paper (Whatman 3MM) below as padding. Place the autoradiogram on top of the gel and align the marks on the X-ray film with the corresponding marks on the dried gel. Hold the gel and the X-ray film in place by using several thumbtacks.

3. Cut the specific cDNA band directly through the X-ray film using a sharp razor blade and collect the gel piece into a 1.5 ml microcentrifuge tube that has 120 μl of TE. Several bands can be cut from the gel in the same way. Use a separate razor blade for every band.

4. Let the tubes stand at room temperature for 10 min. Incubate the tubes in a boiling water-bath for 10 min. Allow the tubes to come to room temperature and centrifuge for 10 min in an Eppendorf centrifuge at top speed to remove any particulate gel material.

5. Collect the eluate carefully into a fresh tube (at this stage the volume of the eluate will be about 100 μl as filter paper absorbs some TE).

6. Add 1 μl (20 μg) of glycogen as carrier and precipitate the cDNA from the eluate by adding 0.1 vol. of 3 M NaOAc pH 6.0 and 2.5 vol. of ethanol. Keep the tubes at −20°C.

7. Collect the precipitate by centrifuging the samples for 15 min and carefully remove the ethanol. Wash the pellet by layering 200 μl of 70% ethanol without disturbing the pellet. Centrifuge again, remove the ethanol, and let the pellet dry at room temperature.

8. Dissolve the pellets in 10 μl of water and centrifuge at top speed for 10 min to remove any insoluble material. 2 μl of this sample is enough as template for PCR amplification.

Protocol 14

Purification of the gel eluted cDNA: PCR

Reagents

- See *Protocol 8*

Protocol 14 continued

Method

1. Use 2 μl of gel eluted cDNA as template and add 2 μl of 2 μM stock solution of the specific 3′ anchored (or heel) primer in a 0.5 ml thin-wall PCR tube. Add 16 μl of the reaction master mix and overlay with mineral oil.

2. The reaction master mix is prepared by aliquoting the following components per tube:
   - 10 × GeneAmp PCR buffer — 2 μl
   - 2 mM dNTPs — 2 μl
   - ^{32}P-labelled primer FA-1 — 2 μl
   - Ampli*Taq* Gold[a] (5 U/μl) — 0.7 U/reaction
   - Water — 10 μl
   - Adjust the final volume to — 16 μl

3. The PCR conditions are as follows.
   (a) Stage I: one cycle of 94 °C for 12 min.
   (b) Stage II: 30 cycles of 94 °C for 30 sec; 60 °C for 2 min; 72 °C for 1 min.

4. Alternatively a two temperature PCR can also be used with the following conditions.[b]
   (a) Stage I: one cycle of 94 °C for 12 min.
   (b) Stage II: 30 cycles of 94 °C for 30 sec; 68 °C for 2 min 30 sec.

[a] It is important to add the enzyme last, after adjusting the volume with required amount of water and gentle mixing.

[b] Both conditions work well. However when heel primers are used then it is better to anneal at 60 °C followed by extension at 72 °C as indicated above.

Protocol 15

Sequencing gel analysis and recovery of the cDNA bands

Reagents

- See *Protocol 11*
- See *Protocol 13*

Method

1. Take 2 μl of the PCR sample, mix with 2 μl of the stop solution, heat denature the sample, and analyse it on a sequencing gel exactly as described earlier (see *Protocol 11*).

2. Dry the gel on a Bio-Rad gel dryer and process it for autoradiography.

3. Expose the gel to a BioMax MR X-ray film for 1 h at room temperature.

4. At this stage we can see intense bands and several faint contaminating bands. The

desired cDNA band can be identified by its position relative to the molecular size markers and in comparison with the original autoradiogram.

5. Excise the gel band and elute the cDNA as described earlier (see *Protocol 13*).

6. Dissolve the cDNA in 10 μl of water.

We need to amplify enough cDNA so that it can be used either for direct sequencing or for cloning of the PCR product.

Protocol 16

PCR amplification of the cDNA with unlabelled primers for sequencing

Reagents

• See *Protocol 8*

Method

1. Set up a 50 μl PCR reaction with 5 μl of the gel eluted cDNA as the template.

  • 10 × GeneAmp PCR buffer   5 μl
  • 2 mM dNTPs   5 μl
  • Primer FA-1 (25 ng/μl)   5 μl
  • 3′ anchored primer or heel primer (2 μM stock)   5 μl
  • AmpliTaq Gold[a] (5 U/μl)   1.5 U/reaction
  • Adjust the final volume to 50 μl with nuclease-free water

2. Overlay the sample with mineral oil and start the PCR reaction.

3. PCR conditions are exactly as described earlier (see *Protocol 14*).

4. At the end of the PCR reaction, the samples can be processed for sequencing as described in Section 2.7.

[a] Always add the enzyme last after adjusting the volume with water and gentle mixing.

Sometimes a specific cDNA band recovered from the display gel does not get amplified in the subsequent PCR steps. This could be due to a variety of reasons including its size and GC content. The original cDNA template is a complex mixture, and PCR amplification of such a mixture behaves differently in comparison to an isolated cDNA band. These problems are more frequent when we attempt to amplify larger size bands. However there are instances where a given band does not get amplified though the size is small. Under these conditions such cDNA bands can be amplified by a modified PCR method developed in our laboratory (24).

Protocol 17

Modified PCR approach for difficult templates

Reagents

- 10 × LA-PCR buffer: 200 mM Tris–HCl pH 9.0, 160 mM ammonium sulfate, 25 mM MgCl$_2$
- 5 M Betaine: dissolve 13.5 g Betaine monohydrate (M$_r$ 135.16, Fluka Biochemica) in autoclaved water in a final volume of 20 ml, and filter the solution using a disposable filtration unit
- 7.5 mg/ml bovine serum albumin (nuclease-free) (New England Biolabs)
- LA-16 enzyme (24): mix Klentaq 1 DNA polymerase (25 U/μl, Ab Peptides, Inc.) and native *Pfu* DNA polymerase (2.5 U/μl, Stratagene) gently on ice, at a ratio of 15:1 (v/v)
- Nujol mineral oil (Perkin Elmer)

Method

1. Take a 0.5 ml thin-wall PCR tube, add 5 μl of gel eluted cDNA template and 5 μl of 2 μM stock solution of specific N1N2 anchored 3′ primer (or heel primer).

2. Add 40 μl of the reaction master mix without dNTPs, overlay with mineral oil, and do a hot start PCR as described below, by adding 5 μl of 2 mM dNTPs for a 50 μl reaction.

3. The reaction master mix contains the following components per tube:
    - 10 × LA-PCR buffer   5 μl
    - 5 M Betaine   10 μl
    - 7.5 mg/ml BSA   1 μl
    - Primer FA-1 (25 ng/μl)   5 μl
    - Water   18.8 μl
    - Enzyme LA-16   0.2 μl
    - Final volume   40 μl

4. PCR conditions.

    (a) Stage I: one cycle of 94 °C for 1 min. Then 80 °C for 10 sec.

    When the tube temperature reaches 80 °C pause the machine at this temperature, add 5 μl of 2 mM dNTP stock solution to each tube, and continue the PCR with the following cycling conditions.

    (b) Stage II: 35–40 cycles of 94 °C for 30 sec; 60 °C for 2 min; 68 °C for 1 min.

5. Alternatively the two temperature PCR can also be done as described earlier (see *Protocol 14*).

6. This modified PCR step can be used with the ^{32}P-labelled primer FA-1 as the first step after recovering the bands from the display gel and/or at the level of amplification with cold primers for agarose gel electrophoresis and sequencing.

2.7 Processing of the PCR amplified samples for sequencing

2.7.1 Direct sequencing of PCR products

Direct sequencing of the purified PCR product is easy and generally results in a good quality sequence. However it is possible that a given PCR product may actually be a mixture of more than one species of cDNA with the same size and hence may result in an unreadable sequence. Under such conditions the specific cDNA band recovered from the gel can be PCR amplified and the products cloned into a suitable vector. While this approach results in very good quality sequence, it is important to sequence the DNA from a significant number of colonies to ensure that a given sequence is represented in statistically significant number of colonies in order to assign its specific position on the display gel. If the sequence obtained corresponds to one represented in the database, then we can analyse the sequence for the N1N2 and the restriction enzyme site that was used in generating the display gel. We can confirm and predict the exact position on the gel and this is very helpful. In general we prefer to do direct sequencing of the PCR products, and only if a given PCR product results in an unreadable sequence do we clone it for sequencing. We use the 5' adapter oligonucleotide (FA-1) as the primer for direct sequencing of the PCR product. This helps in identification of the poly(A) signal and poly(A) sequence of the cDNA band, so that we can evaluate the authenticity of the message.

Protocol 18

Direct sequencing of PCR products

Reagents

- Shrimp alkaline phosphatase (1 U/μl) (Boehringer Mannheim)
- Exonuclease I (10 U/μl) (United States Biochemical (USB) Corporation)
- Tris, acetate, EDTA buffer (TAE): 40 mM Tris–acetate, 1 mM EDTA pH 8.3
- Seakem GTG agarose (FMC Corporation)

Method

1. Prepare a 2% agarose gel in 1 × TAE buffer (20). Analyse 10 μl of the 50 μl PCR product on this gel to ensure that the PCR product is of the right size and that there are no co-amplified contaminating bands (see *Figure 5*).

2. If the product is pure enough and of the expected size, the sample can be processed for sequencing by following *Protocol I* (25). If the sample has some contaminating bands then the specific DNA band can be excised from the gel and prepared for sequencing by following *Protocol II*.

Protocol I

1. Prepare the enzyme master mix in a 0.5 ml microcentrifuge tube (Low Adhesion Microcentrifuge tubes) on ice, by mixing shrimp alkaline phosphatase (1 U/μl) and

Protocol 18 continued

exonuclease I (10 U/μl) at a ratio of 10:1 (v/v). At this stage we will have in the master mix equal amounts of both the enzymes with respect to total number units (10 U: 10 U).

2. Take a 0.5 ml microcentrifuge tube (Low Adhesion Microcentrifuge tubes) on ice and aliquot 5 μl of the PCR amplified product. Add 2 μl of the above enzyme mixture, and incubate the tubes for 15 min at 37°C. Following incubation, heat inactivate the enzymes for 15 min at 80°C. Dilute the samples to approx. 10 ng DNA for every 100 bp, in a final volume of 15 μl with water. Submit the samples for automated sequencing using the 5' adapter oligonucleotide (FA-1) as the primer.

Protocol II

1. Analyse the entire PCR product by electrophoresis on a 2% agarose gel (20). Stain the gel with ethidium bromide to visualize the DNA.

2. Excise the specific band and extract the DNA from agarose gel using a Qiagen Gel Extraction Kit.

3. The eluted DNA can either be used directly for sequencing with the 5' adapter oligonucleotide (FA-1) as the primer, or can be used for cloning into a plasmid vector.

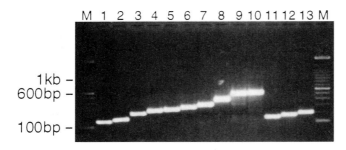

Figure 5 Agarose gel electrophoresis analysis of the PCR amplified cDNA fragments recovered from a differential display gel. Lanes 1–13 represent various cDNA bands recovered from a display gel and analysed as described in the text.

2.7.2 Sequencing by cloning of the gel purified DNA into a plasmid vector

For cloning the PCR products into plasmid vectors, we used two different cloning systems. The pCR-Script cloning method requires (see *Protocol I*) that the ends of the PCR product be polished using *Pfu* DNA polymerase. The agarose gel purified DNA can be polished according to the supplier's protocol. However it is very convenient to polish the DNA immediately following the PCR by adding the *Pfu* DNA polymerase to the reaction and continuing the incubation at 72°C for 30 min (26). Alternatively the pGEM-T easy cloning method does not require (see

Protocol II) the ends of the PCR products to be polished. Both methods worked well.

Protocol 19

Sequencing by cloning of the gel purified DNA into a plasmid vector

Reagents

- pCR-Script Cloning Kit (Stratagene)
- pGEM-T easy Vector system (Promega)

Protocol I

1. Following the PCR reaction take 25 μl of the sample into a fresh PCR tube.

2. Add 1 μl (0.5 U/μl) of *Pfu* DNA polymerase, overlay with mineral oil, and continue incubation at 72 °C for 30 min.

3. At the end analyse the PCR product on a 2% agarose gel and recover the DNA from the specific band as described above using a Qiagen Gel Extraction Kit.

4. The eluted DNA in a final volume of 100 μl (50 μl each, two times for efficient recovery) is then ethanol precipitated in the presence of 0.1 vol. of 3 M sodium acetate.

5. Recover the precipitate by centrifugation at top speed in an Eppendorf centrifuge, wash the pellet once with 70% ethanol, and air dry the pellet by leaving tubes at room temperature.

6. Dissolve the DNA in 10 μl of water. Use about 2–3 μl of the DNA sample for cloning it into pCR-Script vector system. Follow the rest of the steps exactly as suggested by the supplier.

Protocol II

1. Analyse the PCR product directly on a 2% agarose gel and excise the specific DNA band from the gel.

2. Extract the DNA from agarose using a Qiagen Gel Extraction Kit and precipitate with ethanol as described in *Protocol I*.

3. Collect the precipitate by centrifugation, wash once with 70% ethanol, and air dry the pellet.

4. Dissolve the pellet in 10 μl of water and use about 2–3 μl of the DNA sample for cloning into pGEM-T easy Vector system by following the supplier's protocol.

2.8 Data documentation

The autoradiograms are scanned by using a UMAX-Mirage D-16L X-ray film scanner connected to a personal computer for analysis and documentation. Quantitative information on the relative band intensities from the dried sequencing gel can be obtained by PhosphorImager analysis using a Molecular Dynamics PhosphorImager with imagequant software.

3 Discussion

As shown in *Figure 3*, neutrophils exhibit distinct changes in the pattern of gene expression in response to a biologically relevant stimulus such as interaction with *E. coli*. A part of the gel picture in *Figure 3* is enlarged to show these changes more clearly (*Figure 4*). The mechanistic aspects of these changes can be addressed by a variety of ways. It is possible to survey the entire cell for all the changes in the message levels to find out which of these changes are meaningful to the biology of this interaction. We are in the process of such analysis on neutrophils and an investigation on the biological relevance of such changes in response to the interaction with pathogenic bacteria (27).

An average cell may contain mRNAs from perhaps 15000 genes. A rough estimate of the number of gel display lanes necessary to obtain some representation of over 95% of these genes is as follows. Insert sizes from about 75–500 nucleotides give bands on a display gel that are well enough resolved for sequencing. A restriction enzyme with a six nucleotide recognition site randomly distributed along mRNA sequences and of average base composition might cut once every 4500 bases. Roughly 9.5% of these sites will fall between 50–500 nucleotides from an arbitrarily defined site, in this case the poly(A) addition site. Therefore 10% of the mRNAs or somewhat less than 1500 mRNAs will on average produce a band in the resolved range with a single enzyme. According to the Poisson distribution it would take about 30 enzymes to exceed 95% coverage of the mRNAs.

If we assume approximately equal representation of bands in the lanes for each N1N2 then this will give about 150 bands per lane, distributed over somewhat more than 400 positions. If we presume that because of imprecision in poly(A) addition or priming there is on average two bands of scoreable intensity per mRNA this would give 300 bands per lane. Using a Poisson distribution approximately 15% of these might overlay another band so one-sixth of the 9.5% would not have any band resolved from all others on a gel. With these very rough simplifications, one enzyme would then display as resolved bands about 8% of the mRNA, so that an average of a single hit per mRNA would require 12 enzymes, and 95% coverage would require more than 30 enzymes. However more prominent bands can be resolved and analysed even if they are substantially longer than 500 nucleotides, perhaps up to 800 nucleotides. For these products the Poisson distribution predicts that about 18 enzymes would give 95% coverage.

This present protocol results in readily reproducible cDNA display patterns. We also tested the same cDNA sample at different times and the results were consistent and reproducible. It is possible to detect visually differences of about 1.5-fold in the relative intensities for the bands of intermediate intensities. Quantitative estimation of the relative amounts of material present in different bands by use of PhosphorImager suggests that the amount of radioactive cDNA in perceptible bands can vary by nearly four orders of magnitude. Our studies indicate that the most abundant mRNA species are not likely to represent more than 1–2% of the total cellular message population. Based on these factors this approach is able to detect bands corresponding to mRNAs present at less than one part per hundred thousand of total message, which corresponds to approximately one mRNA molecule per cell for a typical cell.

Use of this approach to measure the absolute amounts of individual mRNAs is inaccurate due to a number of factors, such as:

(a) The 'normalization' effect on the most abundant cDNA species, in which self-annealing competes with primer annealing to cDNA strands, decreasing the amplification probability per cycle of PCR (28).

(b) The PCR amplification of longer cDNAs is less efficient as compared to the shorter cDNAs.

(c) It is also possible that the amplification efficiency may vary for cDNAs of similar length because of base composition and/or other undefined factors. Although it is important to keep these factors in mind while using this approach, when examining shifts in the ratio of cDNA bands as a function of changing physiological states of the cell, the benefits outweigh the limitations.

The present method is relatively efficient in terms of sequencing effort. Several enzymes could cut the same cDNA and hence cause some redundancy although in principle, this could be anticipated once the first sequence was obtained. The use of alternative polyadenylation signals and/or the potential inaccuracies at the polyadenylation site of the mRNA could also contribute to redundant bands. We have seen multiple cases attributable to these causes. In addition, there is a possibility of primer slippage during reverse transcriptase priming producing poly(A) that is either copied by an oligo(dT) primer with a mismatch at one of the two 3′ terminal bases, or because of looping out of an internal mismatch base. These phenomena are seen particularly with most abundant mRNA species, and presumably contribute to the background in display gels.

With any cDNA preparation method that uses oligo(dT) priming of cDNA from total cellular RNA rather than cytoplasmic RNA, there is a potential risk of generating at least some of the products from the intranuclear RNA species that include incompletely spliced products and even intergenic transcripts. Priming might also occur at poly(A)-rich sequences that are often associated with interspersed repetitive sequences such as Alu elements (29, 30). The finding of a matching sequence in the EST databases does not give absolute assurance that the sequence is not derived from one of the above artefacts, due to the common features of the cDNA preparation methods.

Interpretations of the display gel bands that correspond to repetitive sequences are problematic. Although a certain fraction of these sequences do represent 3′ ends of the legitimate mRNAs, a substantial fraction of the transcript fragments that contain only repetitive sequences lack appropriately placed definitive polyadenylation signal relative to the apparent polyadenylation site. Sometimes the repetitive sequence is identical over hundreds of bases with an EST sequence in the database making it probable, although not certain, that the cDNA clone is available and could be used to obtain unique sequences for the mRNA. In some cases this can be a difficult call. The repetitive sequence bands that do not exactly match database entries represent a different problem. These sequences cannot be used to probe cDNA libraries because of their repetitiveness and their evaluation would require either relatively intense work or more genomic sequence.

Even without considering all the repetitive sequence bands, about 15% of the bands we obtained did not match any database entries. Some of these had atypical polyadenylation signals. Although this does not rule out that these sequences may correspond to mRNAs, this does raise some caution. However a good number of sequences had adequate or perfect polyadenylation signal sequences at appropriate positions. This suggests that, even for a cell such as neutrophil, that may be present at some level in many tissues, the databases are not entirely complete. The cDNA display methods will remain useful in proportion to the frequency of such uncatalogued transcripts, even if array methods for oligonucleotide hybridization emerge that are able to conveniently detect all known ESTs from a species.

Acknowledgements

The authors gratefully acknowledge Dr Jon Goguen and Dr Nancy Hoe for their help and suggestions on experiments involving the use of bacteria. We also thank Constance Whitney, Padmakumari Yerramilli for their excellent technical help.

References

1. Oliver, S. G., Winson, M. K., Kell, D. B., and Baganz, F. (1998). *Trends Biotechnol.*, **16**, 373.
2. Mewes, H. W., Albermann, K., Bahr, M., Frishman, D., Gleissner, A., Hani, J., *et al.* (1997). *Nature*, **387**, 737.
3. Goffeau, A., Barrell, B. G., Bussey, H., Davis, R. W., Dujon, B., Fledmann, H., *et al.* (1996). *Science*, **274**, 563.
4. Cho, R. J., Campbell, M. J., Winzeler, E. A., Steinmetz, L., Conway, A., Wodicka, L., *et al.* (1998). *Mol. Cell*, **2**, 65.
5. Lashkari, D. A., DeRisi, J. L., McCusker, J. H., Namath, A. F., Gentile, C., Hwang, S. Y., *et al.* (1997). *Proc. Natl. Acad. Sci. USA*, **94**, 13057.
6. The NCBI web page at: `http://www.ncbi.nlm.nih.gov`
7. Proudfoot, N. J. (1996). *Cell*, **87**, 779.
8. Proudfoot, N. J. (1997). *Trends Genet.*, **13**, 430.
9. Schena, M., Shalon, D., Davis, R. W., and Brown, P. O. (1995). *Science*, **270**, 467.
10. Shalon, D., Smith, S. J., and Brown, P. O. (1996). *Genome Res.*, **6**, 639.
11. Lockhart, D. J., Dong, H., Byrne, M. C., Follettie, M. T., Gallo, M. V., Chee, M. S., *et al.* (1996). *Nature Biotechnol.*, **14**, 1675.
12. Velculescu, V. E., Zhang, L., Vogelstein, B., and Kinzler, K. W. (1995). *Science*, **270**, 484.
13. Liang, P. and Pardee, A. B. (1992). *Science*, **257**, 967.
14. Liang, P. and Pardee, A. B. (1995). *Curr. Opin. Immunol.*, **7**, 274.
15. Prashar, Y. and Weissman, S. M. (1996). *Proc. Natl. Acad. Sci. USA*, **93**, 659.

16. Subrahmanyam, Y. V. B. K., Baskaran, N., Newburger, P. E., and Weissman, S. M. (1999). In *Methods in enzymology*, 303, 272. (Sherman, M. Weissman, Ed), Academic Press, San Diego, CA.

17. *SuperScript Choice system for cDNA synthesis:* Instruction manual from Gibco BRL.

18. Gubler, U. and Hoffman, B. J. (1983). *Gene*, **25**, 263.

19. Strohman, R. C., Moss, P. S., Micou-Eastwood, J., Spector, D., Przybyla, A., and Paterson, B. (1977). *Cell*, **10**, 265.

20. Sambrook, J., Fritsch, E. F., and Maniatis, T. (ed.) (1989). *Molecular cloning: a laboratory manual*, 2nd edn. Cold Spring Harbor Laboratory Press, NY.

21. Liu, M., Subrahmanyam, Y. V. B. K., and Baskaran, N. (1999). In *Methods in enzymology*, Vol. 303, 45. (Sherman, M. Weissman, Ed), Academic Press, San Diego, CA.

22. Carninci, P., Nishiyama, Y., Westover, A., Itoh, M., Nagaoka, S., Sasaki, N., *et al.* (1998). *Proc. Natl. Acad. Sci. USA*, **95**, 520.

23. Mizuno, Y., Carninci, P., Okazaki, Y., Tateno, M., Kawai, J., Amanuma, H., *et al.* (1999). *Nucleic Acids Res.*, **27**, 1345.

24. Baskaran, N., Kandpal, R. P., Bhargava, A. K., Glynn, M. W., Bale, A., and Weissman, S. M. (1996). *Genome Res.*, **6**, 638.

25. Werle, E., Schneider, C., Renner, M., Volker, M., and Fiehn, W. (1994). *Nucleic Acids Res.*, **22**, 4354.

26. Dieffenbach, C. W. and Dveksler, G. S. (1995). *PCR primer: a laboratory manual*. Cold Spring Harbor Laboratory Press, NY.

27. Subrahmanyam, Y. V. B. K., Shigeru Yamaga, M.D., Yatindra Prashar, Helen Lee, Nancy Hoe, Jon Goguen, Peter E. Newburger, M.D., and Sherman M. Weissman, M.D. (*1999*) (*Manuscript in preparation*).

28. Mathieu-Daude, F., Welsh, J., Vogt, T., and McClelland, M. (1996). *Nucleic Acids Res.*, **24**, 2080.

29. Schmid, C. W. (1996). *Prog. Nucleic Acid Res. Mol. Biol.*, **53**, 283.

30. Weiner, A. M., Deininger, P. L., and Efstratiadis, A. (1986). *Annu. Rev. Biochem.*, **55**, 631.

Chapter 7

Cloning of differentially expressed brain cDNAs

EILEEN M. DENOVAN-WRIGHT, KRISTA L. GILBY,
SUSAN E. HOWLETT, and HAROLD A. ROBERTSON
Department of Pharmacology, Dalhousie University, Sir Charles Tupper Medical
Building, Halifax, Nova Scotia, B3H 4H7, Canada

1 Introduction

The ability of cells to alter gene expression in response to stimuli or during development and ageing underlies the plasticity of the central nervous system. Although there are multiple levels of control of gene expression, induction or repression of specific genes is an important first step in the regulation of the function of the CNS. Therefore, molecular screening and selection techniques based on comparing mRNA populations can reveal important differences in gene expression between and among brain tissues.

Differential display is a PCR-based screening method (for an overview see 1) that is especially useful to study changes in steady-state levels of mRNA in heterogeneous tissue such as brain for several reasons (1). First, this screening technique can be used to identify novel brain mRNAs or previously described mRNAs whose relative expression levels are altered as a result of a physiological change, a disorder, a disease state, or the administration of pharmacological agents. Secondly, it is possible to simultaneously compare multiple tissues, treatments, and time points (up to 60 comparisons per experiment and primer set) using small amounts of tissue from a number of individual animals for the original isolation of RNA. These features distinguish differential display from methods that require prior information about specific gene expression and from methods that are used to screen or select cDNAs from only two states. Moreover, differential display can be easily established in any laboratory that has basic molecular biology equipment and supplies.

There are a number of differential display protocols that are modifications of the original methods described by Liang and Pardee (2, 3). The differential display method presented here is based on the commercially available Clontech Delta Fingerprinting™ protocol (4). The 5′ primers (designated P1–10) each contain 16 nucleotides at their 5′ end that correspond to the T3 polymerase recognition sequence. Following the T3 polymerase binding sequence are nine nucleotides that are involved in the initial priming of the cDNA during the first few low temperature annealing reactions. The 3′ primers (designated T1–9) contain the recognition sequence for T7 polymerase followed by nine T residues and two nucleotides that are either A, C, or G. These nine T primers, therefore, can anneal with the poly(A) sequence of the cDNA generated after the first linear amplification using the upstream 5′ primer. The 3′ terminal two nucleotides of a given primer effectively reduce the complexity of the PCR fragments by one-twelfth. In order to simplify the screening of potentially differentially expressed bands, we have developed a method to use mixed hybridization probes derived from the differential display gels to probe a Northern blot prior to cloning and sequencing the PCR products. After hybridization, single-stranded DNA probes that are differentially expressed are cloned and sequenced. Therefore, even if the potentially differentially displayed cDNA fragment cut from the denaturing gel contains more than one specific product, only differentially expressed cDNA fragments are cloned. This step reduces the number of Northern hybridization and cloning reactions and eliminates the possibility that differentially expressed cDNAs are overlooked in the secondary screening analysis. Together, these technical refinements have greatly reduced the number of false positive bands that were observed using earlier differential display protocols.

Methods for routine molecular biology techniques such as gel electrophoresis, cloning, and hybridization can be found elsewhere (5).

2 Differential display protocols

Protocol 1

Preparation of solutions and equipment

Equipment and reagents

- Standard molecular biology equipment and reagents
- Oven
- NaOH
- Diethyl pyrocarbonate (DEPC)

Method

1. Standard procedures to minimize RNase contamination are used for all solutions and glassware (5). Whenever possible, disposable sterile plasticware is used for any of the steps prior to the RNA being converted to single-stranded cDNA.

2. To minimize RNase contamination, soak glassware with 0.1 M NaOH for 1 h at room temperature, followed by extensive rinsing with diethyl pyrocarbonate (DEPC) treated double distilled, filter sterilized H_2O (ddH_2O). The glassware is then sterilized by autoclaving and baked in an oven at 160°C overnight.

3. Non-disposable plasticware such as horizontal gel electrophoresis units and centrifuge tubes are also treated with 0.1 M NaOH and rinsed with DEPC treated ddH_2O. Autoclave and bake disposable tubes and Pipetteman tips in an 80°C oven overnight.

4. Solutions and ddH$_2$O are treated to eliminate RNase activity by adding DEPC to a final concentration of 0.05% (v/v), stirring the solution at room temperature for 12–16 h, and then autoclaving the solutions to eliminate DEPC which will inhibit downstream enzymatic reactions.

Protocol 2
Isolation of total cellular RNA

Equipment and reagents

- Dounce homogenizer or Polytron
- RNase-free equipment and instruments
- Brain tissue
- TRIzol™ reagent (Gibco BRL)
- ddH$_2$O, SDS, ethanol, chloroform, isopropanol

Method

1. Remove the brains and isolate the tissue(s) of interest using sterile, RNase-free equipment and instruments. A sterile disposable Petri plate placed on ice can be used as an RNase-free surface for dissection. The tissue is then flash frozen by immersion in liquid nitrogen and stored in cryovials in liquid nitrogen prior to RNA extraction. As high quality RNA is essential for successful differential display, all tissue from an experimental set is stored in liquid nitrogen and the RNA samples are processed simultaneously to reduce the variability among the different RNA samples.

2. Total RNA is isolated using TRIzol™ reagent. The frozen tissue is added to a Dounce homogenizer or centrifuge tube containing 1.0 ml of TRIzol per 50 mg of tissue. Homogenize the tissue using the Dounce homogenizer or, more conveniently, a Polytron until the sample is uniformly dissociated. Rapid homogenization of the brain tissue in TRIzol minimizes RNA degradation. When processing multiple samples, the Polytron unit can be cleaned between samples using a series of washes including a rinse in ddH$_2$O, 1% (w/v) SDS, ddH$_2$O, 70% (v/v) ethanol, and ddH$_2$O.

3. Add 200 μl of chloroform for each 1.0 ml of TRIzol used for homogenization. Shake vigorously for 15 sec and incubate for 3 min at room temperature. Centrifuge the homogenate at 12 000 g for 15 min at 4 °C.

4. Carefully remove the colourless, upper aqueous phase avoiding the material that has collected at the interface. Place the aqueous phase containing the RNA in a clean microcentrifuge tube and add 0.5 ml of isopropanol per 1 ml of the TRIzol reagent used for homogenization. Usually, after phase separation and centrifugation, 600 μl of aqueous phase is recovered. Mix the solution and allow the RNA to precipitate at room temperature for 15 min. Centrifuge at 12 000 g for 15 min at 4 °C. Remove the supernatant and wash the RNA pellet twice using 1.0 ml of 75% ethanol. The samples are vortexed after the addition of each ethanol wash and centrifuged at 7500 g for 3

min at 4 °C to ensure that the pellet has sedimented before the alcohol is removed. A brief centrifugation will collect any residual ethanol in the tube. The last drops of ethanol can be removed using a 26 gauge needle on a syringe.

5. The RNA is allowed to dry for approximately 10 min at room temperature. Protracted periods of air drying or drying the RNA pellet under vacuum will make the resuspension of the RNA difficult. After drying, the RNA samples are resuspended in 1 μl of ddH$_2$O per 1 mg of tissue sample, and heated at 65 °C for 10 min.

6. Determine the concentration of RNA in the sample spectrophotometrically.

Extraction of tissue with TRIzol produces high quality total cellular RNA with little degradation. Although the 260/280 ratio is lower than expected for pure RNA (typically 1.60–1.65 after the first extraction), the total RNA can be used to produce single-stranded cDNA and high quality Northern blots. However, trace amounts of genomic DNA are sometimes present in the RNA samples. Because the DNA in the samples is a substrate for the DNA-dependent RNA polymerases used in the PCR amplification, it will be amplified during the PCR reactions. The resulting PCR fragments, derived from genomic DNA, can mask the cDNA-derived PCR products leading to apparent differences in gene expression between different experimental groups. For this reason, each RNA sample is treated with DNase to eliminate contaminating DNA prior to reverse transcription of the poly(A) mRNA to single-stranded cDNA. *Figure 1* is an example of DNA

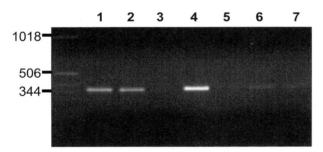

Figure 1 Total RNA samples can contain trace amounts of contaminating genomic DNA as demonstrated by subjecting the RNA samples, before DNase treatment, to PCR amplification using primers that anneal with the genomic DNA sequence of cyclophilin. The PCR products were fractionated on a 1% (w/v) agarose gel in the presence of 0.5 μg/ml ethidium bromide in 0.5 × TBE running buffer. Aliquots of total cellular RNA (lanes 2–7) were used as the substrate for PCR amplification of the 371 bp cyclophilin product according to *Protocol 3*. Each RNA sample (lanes 2–7) was isolated simultaneously using TRIzol™ reagent from an individual rat hippocampus dissected from animals that had received an i.p. injection of 12 mg/kg kainic acid and sacrificed at various times after the kainic acid injection. The substrate for the amplified PCR product fractionated in lane 1 was 1 ng of rat genomic DNA. The presence of the 371 bp band in some, but not all, lanes indicates that various amounts of contaminating genomic DNA are present in these samples. Molecular weight markers (1 kb ladder, Gibco BRL) are shown in the left-hand lane and the size of selected DNA bands in bp is indicated.

contamination in six individual RNA samples that were isolated simultaneously from rat hippocampus. The RNA samples were used as the substrate in PCR reactions that included primers that anneal with cyclophilin genomic or cDNA. Cyclophilin is an ubiquitous, soluble, cytoplasmic 17 kDa protein that binds cyclosporin A (6). The cyclophilin mRNA is expressed in a large number of tissues. The primer combination and PCR conditions given in *Protocol 3* produces a 371 bp fragment from either genomic DNA or cDNA templates. However, as *Taq* polymerase has no affinity for RNA, the amplification of the 371 bp fragment in these samples that had not yet been converted to single-stranded cDNA demonstrates that there is various amounts of contaminating DNA in these six different samples.

Protocol 3

PCR amplification of cyclophilin

Reagents

- RNA (or diluted cDNA)
- 10 × r*Taq* PCR buffer (Pharmacia)
- 5 mM dNTP
- r*Taq* polymerase (Pharmacia)
- Primers

Method

1. In a final volume of 25 μl, mix 1 μl of RNA (or diluted cDNA, see section on quantification of single-stranded cDNA), 2.5 μl of 10 × r*Taq* PCR buffer, 1 μl of 5 mM dNTP, 1 μl of 10 μM 5′ primer (5′-TGG TCA ACC CCA CCG TGT TCT T-3′), 1 μl of 10 μM 3′ primer (5′-GCC ATC CAG CCA CTC AGT CTT G-3′), and 1.0 U of r*Taq* polymerase. The 5′ and 3′ cyclophilin primers anneal with bases 44–65 and 393–414 of the rat cyclophilin cDNA (accession number M19533).

2. The PCR conditions are 94 °C for 1 min, followed by 24 cycles of 94 °C for 1 min, 50 °C for 1 min, and 72 °C for 1 min.

3. Subject an aliquot of the PCR reaction to agarose gel electrophoresis to determine whether or not the RNA samples contain residual genomic DNA. All samples should be treated with DNase if genomic DNA contamination of any of the samples is apparent.

Protocol 4

DNase treatment of total cellular RNA

Reagents

- See *Protocol 2*
- 10 × One-Phor-All Plus buffer (Pharmacia)
- RNasin™ (Promega) (40 U/μl)
- RQ1 RNase-free DNase (Promega) (1 U/μl)

Protocol 4 continued

Method

1. To DNase treat the RNA samples, combine 10 μg of RNA, 5 μl of 10 × One-Phor-All Plus buffer, 1 μl of RNasin™ (40 U/μl), 1 μl of RQ1 RNase-free DNase (1 U/μl), and ddH$_2$O to a final volume of 50 μl.

2. Incubate the samples at 37 °C for 30 min.

3. Extract the DNase treated RNA samples with 100 μl of TRIzol reagent as described in *Protocol 2* to remove the added DNase. The volume of the recovered aqueous phase is approx. 100 μl and, therefore, 84 μl of isopropanol is added to the aqueous phase to precipitate the RNA.

4. After two 70% (v/v) ethanol washes and brief drying, the DNase treated RNA is resuspended in 9 μl of DEPC treated H$_2$O, the concentration of RNA determined spectrophotometrically, and the concentration of RNA in solution adjusted to 500 ng/μl. The 260/280 ratio of the DNase treated RNA is usually between 1.8–1.9 after the second TRIzol extraction.

Protocol 5

Reverse transcription of RNA to single-stranded cDNA

Reagents

- 1 μM oligo(dT) (12–18-mer)
- 5 × first strand cDNA buffer: 250 mM Tris–HCl pH 8.3, 375 mM KCl, 15 mM MgCl$_2$, 50 mM DDT

- 5 mM dNTP
- M-MLV reverse transcriptase (Gibco BRL)

Method

1. To initiate the reverse transcription reactions, combine 2 μg of RNA (4 μl of the 500 ng/μl DNase treated RNA) and 1 μl of 1 μM oligo(dT) (12–18-mer) in an RNase-free microcentrifuge tube.

2. Heat denature the RNA/oligo(dT) mixture at 70 °C for 3 min and then chill on ice for 2 min.

3. Prepare a master mix of first strand cDNA synthesis buffer by combining 2 μl of 5 × first strand cDNA buffer, 2 μl of 5 mM dNTP, and 1 μl of M-MLV reverse transcriptase for each reverse transcriptase reaction to be performed.

4. After the RNA/oligo(dT) mixture has cooled on ice for 2 min, add 5 μl of the first strand cDNA master mix and gently mix the solutions by pipetting.

5. Incubate the reverse transcriptase reactions at 42 °C for 60 min. Although many protocols suggest that reverse transcriptase reactions should be incubated in an air incubator, we have found that a PCR block at 42 °C yields excellent results.

Protocol 5 continued

6. At the end of the 60 min incubation, heat inactivate the reverse transcriptase at 80 °C for 5 min.

7. Dilute the single-stranded cDNA with ddH$_2$O to a final volume of 100 μl. The single-stranded cDNA used for the differential display reactions is stored at −20 °C. It may be convenient to aliquot the dilute cDNA to avoid excessive freezing and thawing of the samples.

PCR amplification of cyclophilin cDNA

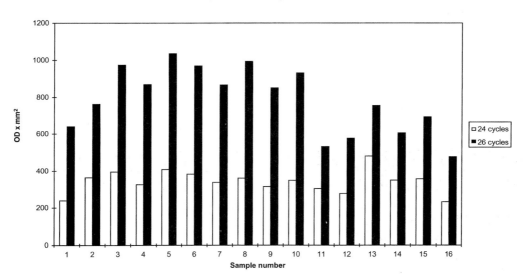

Figure 2 Quantification of the single-stranded cDNA by PCR amplification of cyclophilin. Prior to differential display PCR amplification, relative concentration of reverse transcribed cDNA is estimated by amplifying an aliquot of the dilute cDNA and determining the optical density of a single amplified band fractionated by agarose gel electrophoresis. After 24, 26, and 28 PCR cycles, an aliquot of PCR product was removed and the remaining samples allowed to continue through additional rounds of amplification. Sample number 1–16 correspond to the samples derived from the striatum of 16 individual rats. The variability in the OD among the samples suggests that the concentration of input cDNA is approximately equal. The OD of aliquots from the sample amplification after additional rounds of PCR demonstrates that the PCR components were not limiting during the 24 cycles as the products continue to accumulate at least up to the 26th cycle.

In order to ensure that the final concentration of cDNA is approximately equal in all samples of an experimental set prior to differential display PCR, an aliquot of each sample is subjected to PCR amplification primers that anneal with the cDNA for cyclophilin as described in *Protocol 3*. After 24, 26, and 28 complete PCR cycles, a 5 μl aliquot is removed from each sample. The samples are subjected to agarose gel electrophoresis in the presence of 0.5 μg/ml ethidium bromide and the relative optical density of each ethidium bromide stained band is determined. Samples derived from various cycles are compared in order to ensure that aliquots of each sample are obtained during the exponential amplification of the PCR products. *Figure 2* is an example of PCR amplification of the 371 bp cyclophilin-specific band derived from single-stranded cDNA corresponding to 16 samples of an experimental set. The single-stranded cDNA concentration is approximately equivalent for each sample and, therefore, subsequent differential display banding patterns derived from these single-stranded cDNA samples can be compared.

Protocol 6

Differential display reactions

Reagents

- 10 × cDNA PCR reaction buffer (Clontech)
- 5 mM dNTP
- [α-^{33}P]dATP, 3000 Ci/mmol
- 20 µM primers
- 50 × Advantage™ cDNA polymerase mix (Clontech)

Method

1. Prepare a master mix of reagents required for the PCR amplification of the entire set of single-stranded cDNAs by combining 3.3 µl of ddH$_2$O, 0.5 µl of 10 × cDNA PCR reaction buffer, 0.05 µl of 5 mM dNTP, 0.05 µl of [α-^{33}P]dATP (3000 Ci/mmol), 0.25 µl of 20 µM 5′ primer, 0.25 µl of 20 µM 3′ primer, and 0.1 µl of 50 × Advantage™ cDNA polymerase mix. The 50 × Advantage polymerase mix includes KlenTaq-1 DNA polymerase which because of an N-terminal deletion of the *Taq* polymerase does not have 5′-exonuclease activity, a second DNA-dependent DNA polymerase that has 3′ to 5′ proofreading activity, and TaqStart™ antibody. The TaqStart antibody blocks polymerase activity during the assembling of the PCR reactions. After the initial heat denaturation step, the antibody dissociates from the enzyme and is irreversibly denatured (7). This antibody based 'hot start' for the PCR reactions and the particular enzyme mix in the Advantage KlenTaq polymerase result in highly reproducible differential display fragments with little background.

2. Add 4.5 µl of the differential display master mix to PCR tubes containing 0.5 µl of the dilute single-stranded cDNA. The use of thin-walled PCR tubes and a thermocycler with a heated lid for the differential display reactions makes it unnecessary to use mineral oil to minimize evaporation during the PCR reactions. The reaction volumes can be doubled if evaporation leads to sample to sample variability.

3. The PCR conditions are as follows:
    (1) 94°C for 5 min.
    (2) 40°C for 5 min.
    (3) 68°C for 5 min.
    (4) 94°C for 2 min.
    (5) 40°C for 5 min.

    (6) 68 °C for 5 min.

    (7) Repeat steps 4–6 one time.

    (8) 94 °C for 1 min.

    (9) 60 °C for 1 min.

  (10) 68 °C for 2 min + 4 sec/cycle.

  (11) Repeat steps 8–10 26 times.

  (12) 68 °C for 7 min.

  (13) 0 °C.

Commercially available differential display kits usually include a positive control. For example, the ClonTech Delta RNA Fingerprinting Kit includes cDNA samples derived from fetal and adult liver tissue. Amplification of these cDNA samples using a specific primer combination generates a 380 bp cDNA fragment in the sample derived from fetal but not adult liver (4). While this type of positive control can clearly demonstrate that the differential display reactions are working, it cannot demonstrate that the experimental samples will produce accurate banding patterns. For this reason, it is useful to include an internal positive control that correspond to the identification of a gene that is known to be differentially expressed in the experimental paradigm. For example, immediate-early gene products as *c-fos* have extremely low levels of expression under normal circumstances but are induced by a large number of stimuli (8). It is possible to predict a time, treatment, and tissue in which *c-fos* or other IEGs are expressed. By examining the published cDNA sequence for these or other genes that are known to be differentially expressed under some, but not all, conditions in a particular experiment, a primer combination can be chosen that will amplify a detectable difference between banding pattern for a cDNA fragment of known size. For example, the primer combination P4 and T6 will amplify a band of 553 bp corresponding to the 3′ end of the *c-fos* cDNA. The P4 primer anneals with nucleotides 1564–1571 (GenBank accession number X06769) while the 3′ primer T6 anneals with the cDNA sequence corresponding to the poly(A) tail and terminal two nucleotides of the 3′ UTR of the *c-fos* mRNA. Using this primer combination in the initial differential display reaction can confirm that each of the steps prior to the generation of the differential display PCR products are working successfully and that the original tissue isolated does contain transcripts that are truly differentially expressed. Because of the high degree of reproducibility between different experiments using the same primer combinations and the standard electrophoresis conditions, banding patterns are highly consistent from one set of samples to the next.

We employ a Genomyx LR® Sequencer that uses a buffer gradient and a 6% acrylamide, 8 M urea gel. The electrophoresis conditions are as follows: 50 °C, 3000 V, 100 W, 5 h. The advantage of using this sequencing apparatus is that constant temperature is maintained independently of voltage during electrophoresis. As such, the electrophoretic runs are generally free of anomalies due to local temperature differences in the gel and there is little variability in electrophoretic mobility of bands from one gel to the next. Also the gel can be easily dried on the glass plate following electrophoresis. However, any standard sequencing apparatus can be used to fractionate the differential display reactions. Instead of drying the gel and rinsing to remove excess urea, the gel is transferred directly to 3MM paper and dried completely. However, as the urea can interfere with reamplification of bands excised from the dried gel it may be necessary to optimize the reamplification reaction. Moreover, the urea in the dried gel is hydroscopic and the dried gel will curl and become sticky if care is not taken to keep it well wrapped and dry.

Protocol 7

Gel electrophoresis

Equipment and reagents

- Denaturing loading dye: 10 mM Na_2 EDTA pH 8.0 in deionized formamide, 0.1% (w/v) xylene cyanol, 0.1% (w/v) bromophenol blue
- Sequencing gel equipment
- Whatman 3MM paper
- X-ray equipment

Method

1. Prepare a 6% polyacrylamide/8 M urea sequencing gel.
2. Following the PCR amplification, add 5 µl of denaturing loading dye to 5 µl of PCR product.
3. Heat the samples at 92 °C for 3 min and chill on ice until ready to load the samples on the sequencing gel.
4. Remove traces of acrylamide and flush the top of the gel thoroughly with running buffer to remove excess urea before positioning the shark-tooth comb creating the sample wells.
5. Load 4 µl of PCR product and run the gel until the xylene cyanol has run off the bottom of the sequencing gel.
6. After electrophoresis, the gel is dried directly on the glass plate only until the urea crystallizes throughout the entire gel surface.
7. Rinse the gel with ddH_2O until the gel is transparent and rehydrated. This rinsing removes excess urea which can interfere with the reamplification of differentially displayed bands that are removed from the gel for further analysis.
8. Transfer the gel to Whatman 3MM paper and dry the gel/filter paper to completion.
9. Expose the differential display gel to autoradiography film at room temperature overnight.
10. Identify bands that correspond to differences between experimental conditions. Avoid bands that are not consistently present or absent from each individual sample

of an experimental condition. Outline the band using opaque tape on the auto-radiogram. Align the gel and autoradiography film carefully. This can be done by aligning the edges of the gel with the edge image on the autoradiogram, by piercing the film and gel with a needle before removing the film for development, or by spotting a mixture of dilute radiolabelled isotope in non-water soluble ink on the dried gel before exposing the gel to film. Invert the aligned film and gel on a light box and outline on the back of the 3MM paper the area of the gel corresponding to the differentially displayed cDNA fragment.

11. Remove the film from the gel to avoid damaging the image. Cut the band from the gel/3MM paper using the pencil mark as a guide and store the band at room temperature in a microcentrifuge tube until ready to proceed to reamplification. The gel and film can be stored flat in a dry place indefinitely and bands removed when necessary.

In order to determine the relative variability among the samples of an experimental set, it is useful to perform differential display reactions using 12–20 primer sets for each experiment before examining all the gels and extracting the bands that match the selection criteria. Moreover, as some primer combinations lead to length differences in particular bands, it is possible to compare banding patterns between experiments using a particular primer combination to determine whether or not a putative difference in banding patterns reflects real differences in the RNA populations before extracting the band and testing it by Northern blot hybridization.

Differences in band thickness can reflect quantitative difference in the initial template concentration between samples. It is also possible that increased band thickness is due to the appearance of two distinct bands that appear to co-migrate under one set of electrophoresis conditions. To resolve such differences, independent PCR reactions are performed and electrophoretically fractionated for a longer period of time to distinguish between closely migrating cDNAs and quantitative differences in the initial template. Such replication of PCR and gel electrophoresis can eliminate a significant number of false positive bands.

It is also possible to eliminate some false positive bands by altering the initial concentration of input cDNA. For example, the chance annealing of the primers at 40°C during the initial PCR cycles generates a pool of PCR products that are exponentially amplified during the high temperature annealing reactions. As such, independent PCR reactions using half or twice the original cDNA can demonstrate that apparent differences are due to differences in the original rounds of amplification. These bands can then be eliminated as candidates for differential gene expression prior to further analysis.

The PCR product derived from the differential display band can contain more than one specific product. These PCR products will not differ appreciably in size and cannot be physically separated. Therefore, if the reamplified material is cloned and individual clones are used to probe a Northern blot, it is possible to miss those clones corresponding to the differentially expressed genes despite a positive result on the differential display gel. For this reason, and because we wanted to devise a method to use several potential bands in a single hybridization reaction, we use the PCR amplified products to make hybridization probes using random priming. Both strands of the cDNA molecule(s) are radiolabelled. The PCR reactions are subjected to agarose gel electrophoresis, the bands are removed from the gel, DNA extracted from the agarose, and spectrophotometrically quantified to make a hybridization probe. *Figure 3* is an example of the complex mixture of cDNA fragments that were reamplified from a single differential display band. After hybridization, the cDNA that hybridized with an mRNA was eluted from the immobilized RNA and cloned.

To prepare a Northern blot, total cellular RNA is electrophoretically separated on a 1% denaturing formaldehyde agarose gel and transferred to Zetaprobe (Bio-Rad) membrane using standard methodology (Sambrook *et al.*, 1989). Prior to hybridization, the membrane containing the fractionated RNA is stained with methylene blue to ensure that RNA transfer is complete and that the relative amount of RNA loaded per lane is comparable. The ratio of the 28S ribosomal

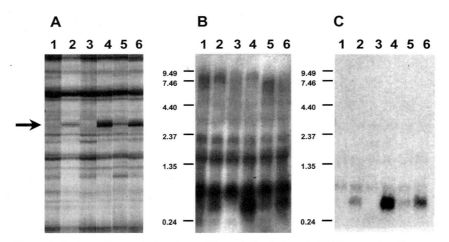

Figure 3 Northern blot analysis using a differentially displayed cDNA fragment as the hybridization probe. Panel A is a portion of the autoradiogram of the fractionated differential display products derived from wild-type (lanes 1, 3, and 5) and cardiomyopathic (lanes 2, 4, and 6) hamsters aged 30 (lanes 1 and 2), 60 (lanes 3 and 4), and 90 days (lanes 5 and 6) using a single differential display primer combination. The band indicated with an *arrow* in panel A was excised from the dried differential display gel, reamplified, and used as a hybridization probe in Northern blot analysis of total cellular RNA derived from the individual animals and fractionated on a 1% agarose, formaldehyde gel (panel B). The complex banding pattern observed in panel B indicates that the reamplified band contained many different cDNA fragments. The hybridizing band present only in the cardiomyopathic hamster RNA samples indicated by an *arrow* in panel B was eluted from the Northern blot, reamplified, cloned, and an individual clone was used to probe another Northern blot (panel C). The size of RNA molecular weight markers (RNA ladder, Gibco BRL) is indicated on the left of the Northern blots in panels B and C.

RNA (rRNA) band to the 18S rRNA band is approximately 2:1. If these rRNAs are not observed as distinct, abundant bands, then it is likely that the RNA has degraded. To stain the filter containing the RNA after high salt transfer from the agarose gel, rinse the membrane for 20 min in 2 × SSC, UV cross-link the RNA to the filter using a Stratalinker (Stratagene), and equilibrate the membrane in 5% (v/v) acetic acid. The filtered methylene blue staining solution contains 0.05% (w/v) methylene blue dissolved in 0.5 M sodium acetate pH 5.2. The membrane with immobilized RNA is stained for 2–10 min and destained in ddH$_2$O until the molecular weight markers and rRNA bands can be clearly visualized. As the binding capacity of the membrane for RNA can be saturated, the relative staining of the RNA is used only as a first approximation of the loading equivalency. After the membrane has been hybridized with potentially differentially expressed cDNA probes, the blot is hybridized with a probe specific for a constitutively expressed gene such as cyclophilin or β-actin. The relative hybridization signals of at least two exposures for the differentially expressed and housekeeping genes are then compared in order to confirm differences in the steady-state levels of the transcript of interest is not due to differences in the amount of RNA fractionated on each lane.

Protocol 8

Reamplification of differentially displayed cDNA fragments

Reagents

- 20 μM primers
- 10 × rTaq buffer (Pharmacia)
- 5 mM dNTP
- rTaq polymerase (5 U/μl)

Method

1. Rehydrate the excised differential display cDNA band in 40 μl of ddH$_2$O by incubating the water containing the gel/paper band at room temperature for 10 min.

2. Heat the sample at 95 °C for 5 min and cool on ice.

3. For the PCR reamplification of the differentially displayed band, combine 10.75 μl of ddH$_2$O, 8.5 μl of the solution from the rehydrated band, 1.25 μl each of the 5′ and 3′ primers (20 μM) used in the original differential display reaction, 2.5 μl of 10 × rTaq buffer, 0.5 μl of 5 mM dNTP, and 0.25 μl of rTaq polymerase (5 U/μl).

4. Subject the samples to the following PCR conditions:

   (1) 94 °C for 1 min.

   (2) 94 °C for 30 sec.

   (3) 58 °C for 30 sec.

   (4) 68 °C for 2 min + 4 sec/cycle.

   (5) Repeat steps 2–4 19 times.

   (6) 68 °C for 7 min.

   (7) 0 °C.

Following the PCR reamplification, subject an aliquot of the product to agarose gel electrophoresis. Although an abundant product is generally observed after 20 PCR cycles, some reamplification reactions must be subjected to more cycles of amplification to obtain sufficient product to make a hybridization probe. Excise the reamplified product from the gel and purify the DNA to remove excess primers using any appropriate reagents and protocol.

Protocol 9

Subcloning of single-stranded cDNAs eluted from Northern blots

Equipment and reagents

- Sephadex G25 spin columns (Pharmacia)
- Pre-hybridization buffer: 50% formamide, 5 × SSC, 1 × Denhardt's reagent, 20 mM sodium phosphate pH 6.8, 0.2% SDS, 5 mM EDTA, 10 μg/ml poly(A), 50 μg/ml sheared salmon sperm DNA, 50 μg/ml yeast RNA
- SSC, SDS
- Whatman 3MM paper

- Hybridization buffer: 50% formamide, 5 × SSC, 10% dextran sulfate, 1 × Denhardt's reagent, 20 mM sodium phosphate pH 6.8, 0.2% SDS, 5 mM EDTA, 10 μg/ml poly(A), 50 μg/ml sheared salmon sperm DNA, 50 μg/ml yeast RNA
- X-ray equipment
- See *Protocol 8*

Method

1. The reamplified PCR fragment is radiolabelled by random hexamer priming using [α-^{32}P]dNTP appropriate to the buffers and conditions employed.

2. Remove excess unincorporated radionucleotides by exclusion chromotography. Sephadex G25 spin columns are convenient for this hybridization probe purification.

3. Pre-hybridize the membrane containing the fractionated RNA for at least 4 h in pre-hybridization buffer at 42 °C.

4. Denature the radiolabelled reamplified PCR product by heating at 95 °C for 5 min and cooling on ice. Add the denatured probe to a final concentration of 2 × 10^6 c.p.m./ml in hybridization buffer. Allow the hybridization to proceed overnight at 42 °C.

5. Wash the blot for four times for 15 min each in 1 × SSC, 0.1% SDS at 55 °C, four times in 0.5 × SSC, 0.1% SDS at 55 °C, twice in 0.25 × SSC, 0.1% SDS at 55 °C, twice in 0.25 × SSC, 0.1% SDS at room temperature, and then expose to autoradiography film at −70 °C for two or more days.

6. Once a true difference in hybridization of the reamplified PCR fragments has been observed on the autoradiogram of the Northern blot, the nylon membrane is placed in direct contact with a piece of 3MM paper on which the outline of the hybridizing band has been drawn from the aligned autoradiogram. The 3MM paper is saturated in ddH$_2$O, carefully aligned to the membrane containing the hybridizing band, and placed in a sealed plastic container in a humidified chamber for 60

min at 37°C in order for the hybridizing band to be eluted from the immobilized RNA and transferred to the 3MM paper.

7. After the 1 h incubation, excise the paper corresponding to the hybridizing band and place it in a sterile microcentrifuge tube.

8. After the addition of 100 μl of ddH$_2$O, a 10 min incubation at room temperature, heating at 95°C for 5 min, and cooling on ice for 2 min, the eluted single-stranded DNA is subjected to PCR amplification using the original differential display primers as described in *Protocol 8*. The reamplification protocol of the bands eluted from Northern blots differs from the original reamplification protocol only in that all of the ddH$_2$O in the reaction is replaced by the eluted DNA solution.

9. The reamplified eluted band is subjected to gel electrophoresis. The gel purified, reamplified band can be cloned into commercially prepared PCR-compatible vectors. Reamplify the eluted band, gel purify the PCR product (ensuring that it is the same size as the fragment used to generate the Northern blot hybridization probe), and set up the ligation reaction without storing the PCR product for maximum cloning efficiency.

10. Identify clones containing recombinant plasmids, isolate plasmid DNA, and sequence the inserts by standard methods.

11. Compare the DNA sequence of the cDNA fragment corresponding to the differentially expressed gene to the data in GenBank in order to identify or infer the identify of the cDNA.

DNA sequence information of potentially differentially expressed cDNAs can be used to generate oligonucleotide probes for *in situ* hybridization. *In situ* hybridization is highly sensitive and gives excellent temporal and anatomical resolution of the distribution of specific transcripts. *In situ* hybridization can also demonstrate the expression of mRNA in areas of the brain not included in the original tissues used to generate RT-PCR fragments. Lastly, the single-stranded cDNA used as the template for differential display and the RNA used in Northern blot analysis are often derived from the same animals. *In situ* hybridization can be used to confirm the differential expression of a transcript in independent samples and eliminate false positives that are the result of microdissection as demonstrated in *Figure 4*.

There are a number of *in situ* hybridization protocols that can be employed. The following is a simple and reliable method based on the protocol described in Wisden *et al.* (9).

Instead of synthesizing oligonucleotide probes, sense and antisense riboprobes can be generated from the DNA-dependent RNA polymerase binding sites on the P and T primers and these riboprobes can be used for *in situ* hybridization. Methods for *in situ* hybridization using riboprobes are described elsewhere. Strand-specific riboprobes can be used in Northern blot analysis to confirm which of the cloned DNA strands is transcribed *in vivo*.

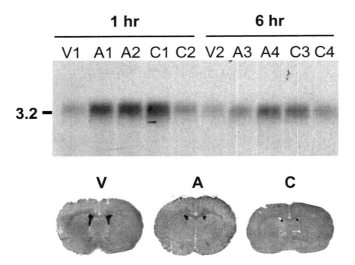

Figure 4 *In situ* hybridization can demonstrate that apparent differences in the differential display gel and Northern blot analysis can be the result of minor contamination of the tissue of interest with tissue derived from different brain areas. Differential display reactions of single-stranded cDNA derived from striatal total RNA of rats that had received saline (V1 and V2), amphetamine (A1, A2, A3, and A4) 5 mg/kg, or cocaine (C1, C2, C3, and C4) 30 mg/kg, and were sacrificed 1 h (V1, A1, A2, A3, and A4) or 6 h (V2, A3, A4, C3, and C4) after i.p. injection were performed. A differentially displayed cDNA fragment that appeared to be expressed in the striatum of amphetamine and cocaine-treated rats was excised from the gel and used as a hybridization blot against fractionated total cellular RNA. A 3.2 kb mRNA species was detected that was apparently more abundant in the drug-treated samples although the expression pattern was variable between some of the time- and treatment-matched individuals. DNA sequence analysis revealed that the cDNA fragment was derived from the mRNA of phosphodiesterase. *In situ* hybridization analysis, however, revealed strong hybridization of a phosphodiesterase-specific oligonucleotide probe to the choriod plexus within the ventricles of vehicle- and drug-treated animals as described previously (10, 11). The *in situ* hybridization, therefore, demonstrated that the apparent difference in expression of the phosphodiesterase I gene was due to contamination of the striatal tissue with tissue contained in the lateral ventricles.

Protocol 10

In situ hybridization

Equipment and reagents

- Cryostat
- SuperFrost™ (Fisher Scientific) slides
- Sephadex G25 spin column (Pharmacia)
- 4% paraformaldehyde in 1 × PBS
- 2 × SSC
- Terminal deoxynucleotidyl transferase (Amersham)

- $[\alpha\text{-}^{33}\text{P}]$dATP 3000 Ci/mmol
- Hybridization buffer: 50% formamide, 5 × SSC, 10% dextran sulfate, 1 × Denhardt's reagent, 20 mM sodium phosphate pH 6.8, 0.2% SDS, 5 mM EDTA, 10 µg/ml poly(A), 50 µg/ml sheared salmon sperm DNA, 50 µg/ml yeast RNA

Method

1. Brains from experimental and control animals are removed and stored at −70°C prior to sectioning. 14 μm sections are cut on a cryostat, thaw-mounted onto SuperFrost™ slides, and stored at −70°C until they are processed for *in situ* hybridization.

2. Frozen sections are allowed to reach room temperature, fixed with 4% paraformaldehyde in 1 × PBS for 5 min, rinsed twice in 1 × PBS for 3 min, once in 2 × SSC for 20 min, and then air dried for approx. 1 h.

3. The oligonucleotides are 3′ end-labelled with [α-^{33}P]dATP for 90 min at 37°C using terminal deoxynucleotidyl transferase for *in situ* hybridization analysis. Prior to use, unincorporated radionucleotides are removed from the labelled probes using a Sephadex G25 spin column.

4. Cover each slide with 200 μl hybridization buffer containing 5×10^6 c.p.m./ml of radiolabelled probe. A piece of Parafilm that has been cut to the exact size of the slide is placed on top of the hybridization buffer. Avoid trapping any bubbles over the brain sections.

5. Incubate the sections overnight at 42°C in a humidified chamber.

6. The Parafilm coverslips are removed from the slides in 1 × SSC and the slides are washed for 15 min four times in 1 × SSC at 55°C, four times in 0.5 × SSC at 55°C, twice in 0.25 × SSC at 55°C, twice in 0.25 × SSC at room temperature, then briefly rinsed in water and air dried.

7. The sections are then exposed to autoradiography film to visualize the hybridizing oligonucleotide probe.

8. Instead of 14 μm fresh frozen brain sections, it is also possible to use brain tissue from animals that have been perfused with paraformaldehyde solution for *in situ* hybridization analysis. The fixed tissue sections are mounted on Superfrost slides, allowed to dry overnight, dehydrated in a graded ethanol series, de-fatted in xylene, and rehydrated in a graded ethanol series prior to being equilibrated in 2 × SSC and air dried for 60 min. Steps 3–7 of the *in situ* hybridization protocol are the same for fresh frozen and fixed brain sections.

References

1. Babity, J. M., Newton, R. A., Guido, M. E., and Robertson, H. A. (1997). In *Methods in molecular biology* (ed. P. Liang and A. B. Pardee), Vol. 85, pp. 285–95. Humana Press Inc., Totowa, NJ.
2. Liang, P. and Pardee, A. B. (1992). *Science*, **257**, 967.
3. Liang, P., Averboukh, L., and Pardee, A. B. (1993). *Nucleic Acids Res.*, **21**, 3269.
4. The Delta RNA Fingerprinting Kit. (1995). *CLONTECHniques*, **X**, 5.
5. Sambrook, J., Fritsch, E. F., and Maniatis, T. (1987). *Molecular cloning: a laboratory manual*, 2nd edn. Cold Spring Harbor Laboratory, Cold Spring Harbor, NY.
6. Iwai, N. and Inagami, T. (1990). *Kidney Int.*, **37**, 1460.

7. Kellogg, D. E., Rybalkin, I., Chen, S., Mukhamedova, N., Vloasik, T., Siebert, P., *et al.* (1994). *BioTechniques*, **16**, 1134.

8. Herrera, D. G. and Robertson, H. A. (1996). *Prog. Neurobiol.*, **50**, 83.

9. Wisden, W., Errington, M. L., Williams, S., Dunnett, S. B., Waters, C., Hitchock, D., *et al.* (1990). *Neuron*, **4**, 603.

10. Narita, M., Goji, J., Nakamura, H., and Sano, K. (1994). *J. Biol. Chem.*, **269**, 28235.

11. Fuss, B., Baba, H., Phan, T., Tuohy, V. K., and Macklin, W. B. (1997). *J. Neurosci.*, **17**, 9095.

Chapter 8

Use of suppression subtractive hybridization strategy to identify differential gene expression in brain ischaemic tolerance

XINKANG WANG and GIORA Z. FEUERSTEIN

Department of Cardiovascular Sciences, DuPont Pharmaceuticals Company, Wilmington, DE 19880, U.S.A.

1 Introduction

Differential gene expression has important functional implications in normal cell growth, differentiation, and in pathophysiological conditions. Ischaemic pre-conditioning (PC) is a phenomenon produced by a short duration of ischaemia that results in a subsequent resistance to severe ischaemic tissue injury (i.e. ischaemic tolerance). This phenomenon has been described in heart and brain (1–3). To identify genes that are specifically regulated in PC, we applied a technique using a PCR-based cDNA subtractive strategy termed *suppression subtractive hybridization* (SSH) (4).

SSH is used to selectively amplify differentially expressed cDNA fragments and simultaneously suppress other DNA species by means of suppression PCR strategy. First, cDNA is synthesized from two types of tissues or cells being compared. The cDNA that contains specific (differentially expressed) genes is referred to as the 'tester' and the control cDNA as the 'driver'. The tester and driver cDNAs are digested with a restriction enzyme (e.g. *Rsa*I) to obtain short blunt-ended molecules. The tester DNA is then divided into two portions, and each is ligated with different 'adapters'. The adapter sequences are specially engineered to prevent undesirable amplification during PCR by means of suppression PCR (5). Secondly, two rounds of hybridization are performed. In the first hybridization, an excess of driver is added to each sample of tester, which results in the significant enrichment of differentially expressed sequences. During the secondary hybridization, the two primary hybridization samples are mixed together without denaturing, which allows the remaining equalized and subtracted tester DNAs to form templates (with two different adapters at both ends) for PCR amplification. Thereafter, using suppression PCR, only differentially expressed sequences are exponentially amplified. The background is reduced and differentially expressed sequences are further enriched following a secondary PCR amplification with a pair of nested primers. These subtracted cDNAs can then be cloned into a vector to generate a SSH cDNA library.

Using the SSH approach, we identified a cDNA clone encoding a tissue in-hibitor of matrix metalloproteinase-1 (TIMP-1); we found that the expression of this gene was up-regulated in the rat cortex by an experimental pre-conditioning procedure (6). TIMP-1 is involved in remodelling of the extracellular matrix by preferential inhibition of matrix metalloproteinases (MMPs) in diverse conditions such as wound healing, scar formation, angiogenesis, and cancer metastasis. Previous studies revealed an increased expression and enzymic activity of MMPs after focal stroke (7), suggesting that TIMP-1 in pre-conditioned cortex may play a role in ischaemic tolerance by inhibiting MMP activity. The study reported here demonstrates the utility of a SSH strategy to discover genes that may con-tribute to ischaemic injury and/or protection; the technique should be valuable to study other disease conditions and aspects of normal development as well.

2 Preparation of RNA for the SSH technique

The first step in the SSH process is to prepare RNA from tissue or cell samples that you wish to compare in order to study differential gene expression in the experimental system.

Protocol 1

RNA preparation

Reagents

- Lysis buffer: 4 M guanidinium thiocynate, 25 mM sodium citrate pH 7.0, 0.5% Sarcosyl. This solution can be stored at room temperature for up to a year. 0.1 M β-mercaptoethanol is added prior to use (after which the solution can be stored safely at room temperature for two months).

- Water-saturated phenol (nucleic acid grade), stored at 4°C
- Chloroform, isoamyl alcohol, isopropanol, all stored at room temperature
- 70% ethanol, stored at −20°C
- 1.67 M sodium acetate pH 4.0, stored at room temperature

Method

1. The pre-conditioning stimulus we used as an experimental paradigm was as follows: a period of 10 min of temporary occlusion of the middle cerebral artery (MCAO) was carried out in spontaneously hypertensive rats to generate focal ischaemic pre-conditioning. This has been described in detail previously (3).

2. The ipsilateral and/or contralateral cerebral cortex was dissected free and immedi-ately frozen in liquid nitrogen and stored at −80°C.

3. Total RNA is prepared by homogenizing the tissues in an acid guanidinium thiocyanate solution and is extracted with phenol and chloroform as described previously (8).

4. Poly(A)$^+$ mRNA is extracted with an oligo(dT) cellulose column from total cellular RNA from the test and control tissues.

3 The suppression subtractive hybridization technique

Protocol 2

Suppression subtractive hybridization

Reagents

- PCR-Select™ cDNA Subtraction Kit (Clontech), stored at −20°C
- PCR-Advantage™ cDNA Polymerase Mix (Clontech), stored at −20°C
- PCR-Select Differential Screening Kit (Clontech), stored at −20°C
- Hybridization buffer: 750 mM NaCl, 50 mM NaH_2PO_4 pH 7.6, 5 mM EDTA, 50% deionized formamide (stored at −70°C), 5 × Denhardt's solution (50 × stock solution stored at 4°C), 2% SDS, 100 μg/ml poly(A), 200 μg/ml boiled salmon sperm DNA, and 10 μl of 0.3 mg/ml of the nested primers

- T/A cloning kit, stored at −20°C
- GeneScreen Plus membrane (DuPont-New England Nuclear)
- Random priming DNA labelling kit (Boehringer Mannheim)
- $[\alpha\text{-}^{33}P]dATP$ (3000 Ci/mmol, Amersham)
- Bio-Spin 6 column (Bio-Rad)
- SuperScript II RNase H⁻ reverse transcriptase and oligo(dT) primers (Gibco BRL)

Method

1. SSH is carried out using the Clontech PCR-Select cDNA Subtraction Kit. 3 μg poly(A)$^+$ mRNA from experimental tissue and from control tissue is used for tester and driver cDNA synthesis, respectively. The double-stranded cDNA is phenol: chloroform extracted and ethanol precipitated. The DNA pellet is resuspended in water and digested with a restriction enzyme (e.g. RsaI), then phenol:chloroform extracted, and resuspended in 5.5 μl H_2O.

2. Adapter 1 and adapter 2R are each separately ligated to 2 μl of a 1:6 dilution of enzyme digested tester DNA at 16°C overnight. Samples are then heated at 70°C for 5 min to inactivate the ligase. Ligation efficiency is analysed prior to the subtractive hybridization.

3. In the first hybridization, 1.5 μl enzyme digested driver cDNA is mixed with 1.5 μl of diluted adapter 1-ligated tester or adapter 2R-ligated tester (i.e. about 18 copies of driver DNA versus 1 copy of tester) in the presence of hybridization buffer, then is covered with mineral oil. The samples are denatured at 98°C for 1.5 min and immediately incubated in a thermal cycler at 68°C for 12 h.

4. In the second hybridization, the two samples from the first hybridization are mixed in the presence of a freshly denatured (1 μl of 1:2 dilution) driver. The sample is incubated at 68°C for 18 h. After adding 200 μl of dilution buffer the sample is incubated for an additional 7 min.

5. The primary PCR is conducted in 25 μl, containing 1 μl of the diluted subtraction mixture, 1 μl PCR primer 1 (10 μM), 10 × PCR reaction buffer, 0.5 μl dNTPs mix

(10 mM), and 50 × Advantage cDNA Polymerase Mix, according to the manufacturer's instructions. The reaction mixture is incubated at 75°C for 5 min to extend the adapters, followed by 94°C for 25 sec in a PCR system (e.g. a Perkin Elmer GeneAmp PCR System 9700). This is followed by 27 cycles at: 94°C for 10 sec, 66°C for 30 sec, 72°C for 1.5 min.

6. The primary PCR mixture is diluted tenfold and 1 μl is used in a secondary PCR with nested primers for primer 1 and primer 2R. The conditions of the reaction are: 94°C for 10 sec, 68°C for 30 sec, 72°C for 1.5 min, for ten cycles.

Protocol 3

Evaluation of subtraction efficiency

Analysis of the subtraction efficiency is carried out using one or both of the following methods.

Method 1

1. PCR amplification of G3PDH for the diluted subtracted cDNA pool versus unsubtracted control cDNA, done according to the manufacturer's instructions. The diluted unsubtracted cDNA is the mix of the ligations of the tester with adapter 1 and adaptor R2 (prior to the overnight incubation).

2. The dilution is 1 μl in 1 ml H_2O. PCR is performed for 33 cycles at: 94°C for 30 sec, 60°C for 30 sec, and 68°C for 2 min.

3. The PCR product is monitored on a 2% agarose gel for an aliquot removed from each reaction after 18, 23, 28, and 33 cycles.

Method 2

1. The subtraction efficiency is evaluated by Southern hybridization of the first and secondary PCR amplification products from the subtracted and unsubtracted DNA to a housekeeping gene (e.g. rpL32) (see refs. 6 and 9).

Protocol 4

Screening of the subtracted cDNA

Equipment and reagents

- 96-well bio-dot apparatus (Bio-Rad)
- T/A cloning kit (Invitrogen)
- PCR-Select Differential Screening Kit (Clontech)
- Pre-hybridization buffer: 5 × SSPE, 50% formamide, 5 × Denhardt's solution, 2% SDS, 200 μg/ml salmon sperm DNA, 100 μg/ml poly(A)

- 0.5 NaOH, 1.5 NaCl
- 0.5 M Tris–HCl
- $[\alpha\text{-}^{32}P]$dATP, $[\alpha\text{-}^{32}P]$dCTP (Amersham)
- QIAquick PCR Purification Kit (Qiagen)
- Superscript II RNase H⁻ reverse transcriptase (Gibco BRL)

Protocol 4 continued

Method

1. The selective secondary PCR products are cloned into a pCR®2.1 vector using a T/A cloning kit. The ligation is transformed into an *E. coli*, IVNαF'.

2. Screening of the subtracted cDNA sample is carried out partially using the PCR-Select Differential Screening Kit against randomly selected bacterial colonies cultured overnight in a 96-well plate and blotted onto the 96-well bio-dot apparatus.

3. The membranes are treated with 0.5 NaOH plus 1.5 NaCl for 4 min, and with 0.5 M Tris–HCl for 4 min. Then the membranes are washed for 30 min in $0.2 \times$ SSC, 0.2% SDS at 63 °C. Pre-hybridization is carried out at 42 °C in pre-hybridization buffer containing 10 μl/ml of 0.3 mg/ml of oligonucleotides corresponding to the nested primers and complementary sequences.

4. Probes were generated by either of the following two methods:

   (a) Random priming method. In this method, the subtracted cDNAs are used as the templates in the presence of DNA polymerase, random primers, and [α-^{32}P]dATP. To reduce the hybridization background, the adapter in the subtracted templates are removed by use of a QIAquick PCR Purification Kit after appropriate restriction enzyme (according to the sites in the adapter sequences) digestion.

   (b) Reverse transcription (RT) method. In this method, poly(A)$^+$ RNA isolated from either the test tissue or control tissue is used as a template, and Superscript II RNase H$^-$ reverse transcriptase is used for the labelling reaction in the presence of both [α-^{32}P]dATP and [α-^{32}P]dCTP according to the manufacturer's instructions.

5. The hybridization is performed overnight with $1–2 \times 10^6$ c.p.m./ml. The membranes are washed at room temperature for 15 min with $2 \times$ SSC, followed by 66 °C with $2 \times$ SSC, 0.5% SDS for 3×15 min, and $0.2 \times$ SSC, 0.1% SDS for 10–15 min. The hybridization results are analysed after autoradiography.

As illustrated in *Figure 1*, a panel of SSH clones was generated in our experiments from cortical samples 24 h after pre-conditioning by subtracting normal cortex, and was analysed by comparative Southern dot-blot hybridization. Noticeably, one clone (indicated with an *arrow* in *Figure 1*) was markedly induced in the pre-conditioned samples compared with the normal controls.

Confirmation of gene expression is a critical step following SSH cloning, in spite of the low incidence of false positive clones compared to other techniques such as mRNA differential display. A variety of techniques can be used for this purpose. Since Northern analysis has been most widely used, it is illustrated here.

As illustrated in *Figure 2*, the temporal expression of our differentially expressed gene (TIMP-1) mRNA following the test pre-conditioning stimulus was examined and showed significant induction over the control situation.

153

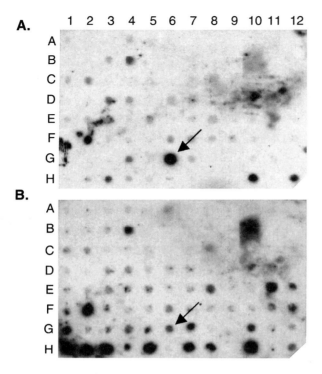

Figure 1 Comparative Southern blot analysis of cDNA clones identified by the suppression subtractive hybridization technique. cDNA clones were generated by the suppression subtractive approach and cultured in a 96-well dish. The bacterial cultures were transferred onto a nylon membrane of a dot-blot apparatus and then analysed by Southern hybridization. (A) Southern hybridization of the membrane using a probe generated from the subtracted PC (24 h) samples. (B) Southern blot analysis of the same membrane with a probe generated from normal cortical samples by means of reverse transcription reaction labelling. The clone indicated with an *arrow* is TIMP-1 as determined by DNA sequencing analysis.

Protocol 5

DNA sequencing and database search

Equipment and reagents

- Double-strand DNA cycle sequencing system (Gibco BRL), store at $-20\,°C$

- $[\alpha\text{-}^{32}P]ATP$ (5000 Ci/mmol, Amersham)

- PNK: T4 polynucleotide kinase at stock concentration of 10 U/μl; store at $-20\,°C$

- $10 \times$ PNK buffer: 700 mM Tris–HCl pH 7.6, 100 mM $MgCl_2$, 50 mM DTT; store at $-20\,°C$

- 6% polyacrylamide/8 M urea sequencing gel

Method

1. DNA sequencing is essential to identify the genes discovered by SSH.

Protocol 5 continued

2. Since the DNA fragments have already been cloned into a pCR®2.1 vector, DNA sequencing can be conveniently carried out using a universal primer.

3. Using the sequencing information, the identity of the differentially expressed genes can be determined by searching a database such as GenBank.

4. If the sequence represents an unknown gene, a cDNA library can be screened using this DNA as a probe to obtain the full-length cDNA clone.

5. As an example, the clone identified in *Figure 1* was sequenced and identified to be TIMP-1 via a GenBank database search.

Protocol 6

Northern blot analysis

Equipment and reagents

- RNA samples, stored at −80 °C
- GeneScreen Plus membrane (DuPont-New England Nuclear)
- Northern hybridization buffer: 5 × SSPE, 50% deionized formamide (stored at −70 °C), 5 × Denhardt's solution (50 × stock solution stored at 4 °C), 2% SDS, 100 μg/ml poly(A), and 200 μg/ml boiled salmon sperm DNA; prepare a fresh solution just prior to use

- 5 × SSPE: 750 mM NaCl, 50 mM NaH_2PO_4 pH 7.6, 5 mM EDTA
- Random priming DNA labelling kit (Boehringer Mannheim)
- [α-^{33}P]dATP (3000 Ci/mmol, Amersham)
- Bio-Spin 6 column (Bio-Rad)
- Washing solution: 2 × SSPE, 2% SDS; store at room temperature

Method

1. For Northern analysis, 30 μg/lane total cellular RNA is electrophoresed through formaldehyde agarose gel and transferred to a GeneScreen Plus membrane.

2. cDNA fragments for the differentially expressed gene and a housekeeping control gene (such as rpL32) are released from the plasmid and gel purified.

3. The cDNA probes are uniformly labelled with [α-^{32}P]dATP (3000 Ci/mmol) using a random priming DNA labelling kit.

4. Hybridization and washing are performed as described in detail previously (6 and 9).

5 Summary

The SSH technique is sensitive and efficient for generating cDNAs highly enriched for differentially expressed genes of both high and low abundance because it combines both PCR amplification (especially the suppression PCR) and cDNA subtraction. To date, other techniques such as mRNA differential display, conventional subtractive library screening, and representational differ-

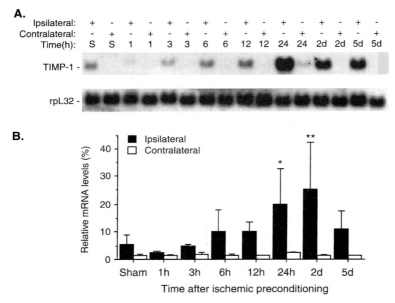

Figure 2 Temporal expression of TIMP-1 mRNA in cortical samples after 10 min MCAO in rats. (A) A representative autoradiograph showing the time course of expression of TIMP-1 mRNA after experimental pre-conditioning. Total cellular RNA (40 μg/lane) was resolved by electrophoresis, transferred to a nylon membrane, and hybridized to the indicated cDNA probe. Ipsilateral and contralateral cortical samples (denoted by +) from individual rats of sham surgery (S; 24 h) or 1, 3, 6, 12, and 24 h, and two and five days after pre-conditioning are illustrated. (B) Quantitative Northern blot data (n = 5) for TIMP-1 mRNA expression after pre-conditioning. The data were generated using PhosphorImager analysis and displayed graphically (mean + standard deviation) after normalizing with rpL32 mRNA signals. Statistical evaluation was performed using five complete sets of cortical samples from each time point using one-way ANOVA followed by a Fisher's protected t test. *$p < 0.05$ and **$p < 0.01$, compared to sham-operated animals.

ential analysis have also been widely used for novel gene discovery. While the mRNA differential display technique (10) is popular because of its simplicity and sensitivity, it generates high incidences of false positives. Although the traditional subtraction approach (9, 10) is reliable, it is biased to detect differential expression of abundant genes only. Representational differential analysis (9, 10, 12, 13) is sensitive, but this technique does not address adequately the problem of the large differences in abundance of individual mRNA species.

While the SSH technique has significant advantages, some disadvantages in using the technique must be acknowledged. The technique is relatively complicated and requires a large measure of technical skill. Because of this, quality control experiments are critical to ensure that the subtraction is successful. In addition, because the nature of the SSH strategy, cDNA clones generated by this technique are small. Therefore, it often requires full-length cDNA cloning after lead identification if the gene is novel.

The application of SSH and other techniques for isolating differentially expressed genes associated with disease processes will no doubt facilitate the discovery of novel therapeutic targets as well as help in the understanding of the molecular mechanisms of disease. It should be pointed out, however, that such gene 'discovery' is only the first step in a long process. Determination of the function of such differentially expressed genes in physiological and pathological processes may be a complicated but necessary procedure to aid in their validation as potential pharmaceutical targets.

References

1. Kitagawa, K., Matsumoto, M., Tagaya, M., Hata, R., Ueda, H., Niinobe, M., *et al.* (1990). *Brain Res.*, **528**, 21.
2. Yellon, D. M. and Baxter, G. F. (1995). *J. Mol. Cell. Cardiol.*, **27**, 1023.
3. Barone, F. C., White, R. F., Spera, P. A., Ellison, J., Currie, R. W., Wang, X. K., *et al.* (1998). *Stroke*, **29**, 1937.
4. Diatchenko, L., Lau, Y. C., Campbell, A. P., Chenchik, A., Moqadam, F., Hung, B., *et al.* (1996). *Proc. Natl. Acad. Sci. USA*, **93**, 6025.
5. Siebert, P. D., Chenchik, A., Kellogg, D. E., Lukyanov, K. A., and Lukyanov, S. A. (1995). *Nucleic Acids Res.*, **23**, 1087.
6. Wang, X. K., Yaish-Ohad, S., Li, X., Barone, F. C., and Feuerstein, G. Z. (1998). *J. Cereb. Blood Flow Metab.* **18**, 1173.
7. Mun-Bryce, S. and Rosenberg, G. A. (1998). *J. Cereb. Blood Flow Metab.*, **18**, 1163.
8. Chomczynski, P. and Sacchi, N. (1987). *Anal. Biochem.* 162, 156.
9. Wang, X. K., Barone, F. C., White, R. F., and Feuerstein, G. Z. (1998). *Stroke*, **29**, 516.
10. Liang, P. and Pardee, A. B. (1992). *Science*, **257**, 967.
11. Herick, S. M., Cohen, D. I., Nielson, E. A., and Davis, M. M. (1984). *Nature*, **308**, 149.
12. Lisitsyn, N., Lisitsyn, N., and Wigler, M. (1993). *Science*, **259**, 946.
13. Hunbank, M. and Schatz, D. G. (1994). *Nucleic Acids Res.*, **22**, 5640.

Chapter 9

Studying gene expression profiles in specialized brain regions by microSAGE

ERNO VREUGDENHIL and NICOLE DATSON
Division of Medical Pharmacology, LACDR, Leiden University, P.O. Box 9503, 2300 RA Leiden, The Netherlands

1 Introduction

The human genome is thought to contain 100 000 genes of which a subset of approximately 15 000 to 20 000 genes is expressed in an individual cell. The set of genes expressed and the stoichiometry of the resulting messenger RNAs, together called a transcriptome, determine the phenotype of a cell, tissue, and whole organism. It is generally accepted that a transcriptome is largely determined by an interplay of hereditary and environmental factors. For example, in the CNS, a challenge from the environment, e.g. a learning or a traumatic experience may lead to an alteration of the transcriptome of target neurons. Thus, transcriptome analysis and subsequent transcriptome comparisons may reveal novel insights in the molecular mechanisms underlying complex processes such learning and memory formation.

Several methods, aimed at determining complete transcriptomes, have become available during the last few years. Among these, serial analysis of gene expression (SAGE) is a very powerful technique to characterize transcriptomes, in particular to identify differentially expressed genes. SAGE is a highly sensitive PCR-based method which yields both qualitative and quantitative information on the composition of a mRNA pool or transcriptome. Since its introduction in 1995 (1), a number of SAGE-related articles have been published (2–5) which have demonstrated the enormous potential of SAGE to detect changes in expression levels of large numbers of genes simultaneously. Already, these studies have led to the identification of a number of new genes with relevance for cancer and development.

SAGE is based on the generation of short approximately 10 bp transcript-specific sequence tags and ligation of the tags to long strings (concatemers), which are subsequently cloned and sequenced (please see *Figure 1* in plate section). The frequency of each tag in the concatemers reflects the original stoichiometry of the individual transcripts in the mRNA pool, allowing quantitative assessment of gene expression. By comparing expression profiles derived from different mRNA sources, differentially expressed genes can be identified. In a single sequence reaction, over 25 tags (cDNAs) can be analysed which gives a 25-fold increase in efficiency compared to expressed sequence tag (EST) sequencing. These short tags are specific enough to uniquely identify each transcript and can be linked to known genes or ESTs present in GenBank, facilitating retrieval of additional sequence of potentially interesting up-regulated or down-regulated tags and their further characterization.

Despite its sensitivity, efficiency, and reliability, SAGE also has a couple of disadvantages. First, large numbers of tags need to be sequenced to enable detection of differences in expression of low abundant transcripts, requiring high-throughput sequencing facilities and robotics. Secondly, SAGE is less useful for analysing gene expression in organisms that are relatively underrepresented in GenBank. Thirdly, SAGE requires a large amount of starting RNA (2.5–5 μg poly(A)$^+$ RNA). The latter point is of particular relevance for gene expression profiling in the central nervous system which consists of many unique, highly specialized, and often small substructures, each with their own specific ex-pression profile. RNA isolated from complex tissues consisting of heterogeneous cell populations—e.g. as is the case in whole brain—will dilute the relative expression profile of the different cell types and thus may mask relevant changes in expression.

In this chapter we describe a modified SAGE protocol, called microSAGE which requires minimal amounts of starting material. MicroSAGE is especially relevant for expression profiling in neuronal tissue, in which obtaining large amounts of homogeneous tissue for RNA isolation is generally an impossible task due to its complex circuitry and high degree of specialization. Using this protocol, we can obtain an expression profile from a single hippocampal punch derived from a 300 μm brain slice, which we estimate to contain at least a factor 5×10^3 less poly(A)$^+$ RNA than is required for the original procedure.

2 Steps in the SAGE procedure

The various steps comprising the microSAGE procedure are outlined in *Figure 2* (please see plate section). Total RNA is isolated from the tissues or cell lines of interest (**1.1**). Via annealing of a biotinylated oligo(dT) primer, the mRNA fraction is immobilized in a streptavidin-coated PCR tube (**1.2 A**). The bound mRNA fraction is converted to double-stranded cDNA (**1.2 B**), which is subsequently digested with a restriction enzyme with a 4 bp recognition site, called anchoring enzyme (AE) (**1.3**). (Note: the 3′ end of the cDNA, from the poly(A) tail to the most 3′ AE site, remains bound to the wall of the streptavidin-coated tube, while the remainder of cDNA fragments are removed by washing.)

Two different linkers (**1.4 A**) are ligated in equal amounts to the cDNA (**1.4 B**). Both linkers contain a recognition site for a type IIS restriction enzyme, called tagging enzyme (TE). Digestion with the TE, which cuts at a defined distance 3′ from the recognition site, results in the release of the linker joined to approximately 10 bp of cDNA, *the tag* (**1.5**). (Note: the remainder of the cDNA remains bound to the wall of the tube. Only the linker with the adjoined tag is cleaved off.)

The released cDNA tags are subsequently blunted (**1.6**) and ligated to ditags (**1.7**). A PCR amplification is performed on the ligated ditags using primers directed against the linker sequences (**1.8**). A limited number of cycles of re-PCR are performed on the amplified ditag to generate sufficient material for the subsequent steps of the procedure (**1.9**). (Note: since all ditags have an equal length there is no length-based PCR bias favouring amplification of a subset of molecules. This ensures that the stoichiometry of the individual tags is not skewed.)

After sufficient quantities of the ditag product have been generated by PCR (**1.10**), the linkers are cleaved off by digestion with the AE (**1.11**), the ditag is isolated from gel (**1.11**), and ligated to long concatemers of ditags (**1.12**). After a size selection (**1.12**), the concatemers are cloned (**1.13**) and sequenced (**1.14**). The raw sequence data is analysed using the special SAGE software, allowing extraction of individual tags from the concatemers, assessment of tag frequency, and comparison with GenBank sequences (**1.15**).

Protocol 1

RNA isolation

In principle any standard method for isolating total RNA can be used, but we recommend using TRIzol.

General remarks

An important factor in the overall success of (micro)SAGE is the RNA quality. To minimize the chance of RNA degradation:

- Wear gloves.

- Use sterile tips and tubes. Incubate all plasticware at 110°C O/N.

- Use RNase-free glassware. Incubate glassware at 180°C O/N.

- All buffers have to be RNase-free. To this end use diethyl pyrocarbonate (DEPC) treated double distilled water (add 2.5 ml DEPC to 2.5 litres of double distilled water, incubate O/N at RT, and autoclave).

- Work on ice.

Equipment and reagents

- Micropestle
- TRIzol (Gibco BRL)
- Chloroform
- 20 mg/ml glycogen (Boehringer Mannheim)

- DEPC treated H_2O
- 3 M NaAc
- Ethanol
- 70% ethanol

A. Punches of rat brain tissue sections

The method described here uses punch material removed from rat brain slices as starting material and is particularly suitable for small quantities of tissue from which only limited

Protocol 1 continued

amounts of RNA can be isolated (please see *Figure 3* in plate section). In principle any tissue source can be used for microSAGE, providing that it is directly frozen at −80 °C or in liquid nitrogen, or immediately used for RNA isolation. If 2.5–5 μg of poly(A)$^+$ RNA can be isolated as starting material we recommend to use the SAGE protocol described by Velculescu *et al.* (1).

1. After decapitation, remove rat brain from the skull and transfer rapidly to iso-pentane on a mixture of dry ice and ethanol. Store at −80 °C until further use.

2. Use a cryostat to prepare thick coronal sections, e.g. 300 μm, at −18 °C.[a]

3. Thaw-mount on poly-L-lysine coated slides and if necessary store at −80 °C.

4. Using a hollow needle (0.3 mm in diameter) chilled by dipping in liquid nitrogen, punch out part of the selected region by the Palkovits method (6).

5. Transfer punch **directly** to a tube containing 20 μl of TRIzol (Gibco BRL).[b]

6. Shake vigorously to disrupt the tissue and store at 4 °C until further use.

B. RNA isolation

1. Add TRIzol to a total volume of 200 μl and homogenize tissue further using a micropestle.

2. Incubate TRIzol solution containing the RNA at RT for 5 min.

3. Add 0.2 ml chloroform/ml TRIzol.

4. Vortex for 15 sec. Let stand at RT for 30 sec, and vortex again for 15 sec.

5. Transfer solution to 2.0 ml Eppendorf tubes.

6. Spin for 15 min at 15 000 g at 4 °C.

C. RNA precipitation

1. After spinning transfer the colourless upper phase to a new Eppendorf tube.

2. Add 0.5 ml isopropanol/ml TRIzol.

3. Add 1 μl glycogen (20 mg/ml) as a carrier.

4. Vortex and incubate 10 min at RT.

5. Spin for 10 min, 12 000 g at 4 °C.

6. Carefully remove supernatant.

7. Wash the RNA pellet by adding 75% ethanol (1 ml ethanol/ml TRIzol).

8. Vortex.

9. Spin 5 min, 7500 g at 4 °C.

10. Carefully remove supernatant.

11. Air dry pellet for 15 min.[c]

12. Dissolve pellet in DEPC treated ddH$_2$O.

13. To store RNA, add 0.1 vol. of 3 M NaAc and 2 vol. of absolute ethanol. Aliquot RNA in portions of 2.5 μg and store at −70°C.

[a] Often, it is difficult to recognize specific brain regions in unfixed or unstained tissue. To solve this, adjacent sections of approx. 75 μm can be prepared and stained with specific markers, e.g. antibodies. These may serve as landmarks to facilitate punching out of the correct region.

[b] Multiple punches may be pooled together for further processing.

[c] The pellet should not be too dry. Do not use SpeedVac to dry the pellet as it will be difficult to dissolve.

Protocol 2

mRNA capture and cDNA synthesis

Reagents

- mRNA Capture Kit (Boehringer Mannheim)
- SuperScript™ II RNase H⁻ reverse transcriptase (200 U/μl; Gibco BRL)
- 0.1 M DTT (supplied with SuperScript)
- 5 × first strand buffer: 250 mM Tris–HCl pH 8.3, 375 mM KCl, 15 mM MgCl$_2$ (supplied with SuperScript)
- dNTPs (10 mM each) (HT Biotechnology)

- 5 × second strand buffer: 100 mM Tris–HCl pH 6.9, 450 mM KCl, 23 mM MgCl$_2$, 0.75 mM β-NAD⁺, 50 mM (NH$_4$)$_2$SO$_4$ (Gibco BRL)
- DNA polymerase I (10 U/μl; Gibco BRL)
- T4 DNA ligase (5 U/μl; Gibco BRL)
- RNase H (1 U/μl; Boehringer Mannheim)
- DEPC treated ddH$_2$O (see *Protocol 1*)
- H$_2$O

A. mRNA capture

1. After RNA precipitation, resuspend the total RNA pellet in 20 μl lysis buffer (mRNA Capture Kit). Dilute the biotinylated oligo(dT)$_{20}$ primer (mRNA Capture Kit) 20 × to a final concentration of 5 pmol/μl.

2. Add 4 μl of the diluted primer to the RNA in lysis buffer.

3. Anneal primer to RNA at 37°C for 5 min.

4. Transfer the RNA to a streptavidin-coated PCR tube (mRNA Capture Kit).

5. Incubate at 37°C for 3 min (in this step the mRNA is immobilized to the wall of the tube by streptavidin-biotin binding).

6. Remove the solution from the tube (contains the non-bound RNA fraction: ribosomal RNA, tRNA, etc.) and wash tube gently 3 × with 50 μl washing solution (mRNA Capture Kit).

Protocol 2 continued

7. Remove the washing solution of the final wash step.

8. Proceed immediately to cDNA synthesis step.

B. cDNA synthesis[a]

1. Rinse tube with captured poly(A)$^+$ RNA once with 50 μl of $1 \times$ first strand buffer, then remove buffer.

2. Replace $1 \times$ first strand buffer with (pipette on ice!):
   - 4 μl of $5 \times$ first strand buffer
   - 2 μl of 0.1 M DTT
   - 1 μl of 10 mM dNTPs
   - 1 μl SuperScript II RT (200 U/μl)
   - 12 μl DEPC H$_2$O
   - Final volume 20 μl

3. Incubate at 42 °C for 2 h (it is convenient to perform the incubation in a PCR machine).

4. Remove reaction mixture from tube and rinse once with 50 μl washing solution (mRNA Capture Kit).

5. Remove washing solution and rinse tube once with 50 μl $1 \times$ second strand buffer.

6. Replace $1 \times$ second strand buffer with:
   - 4 μl of $5 \times$ second strand buffer
   - 0.4 μl of 10 mM dNTPs
   - 1 μl DNA polymerase I (10 U/μl)
   - 0.5 μl T4 DNA ligase (5 U/μl)
   - 0.5 μl RNase H (5 U/μl)
   - 13.6 μl H$_2$O
   - Final volume 20 μl

7. Incubate at 16 °C for 2 h.

8. At this point the double-stranded cDNA can be stored at -20 °C or directly used in the next step (digestion of cDNA with anchoring enzyme).[b]

[a] The oligo(dT)$_{20}$ primer (biotinylated; supplied with mRNA Capture Kit) used to capture the poly(A)$^+$ RNA and immobilize it to the wall of the tube serves directly as a primer for cDNA synthesis. The synthesized cDNA remains bound to the wall of the tube until it is released by digestion with TE.

[b] After synthesis of double-stranded cDNA it is advisable to check the quality by performing RT-PCR. If possible choose primers for a high abundant transcript (housekeeping gene) and also a low abundant transcript known to be expressed in the used RNA source. Failure to detect the transcripts by RT-PCR may be indicative of inefficient cDNA synthesis.

Protocol 3

Digestion of cDNA with anchoring enzyme

Reagents

- 10 U/μl NlaIII[a] (New England Biolabs)
- 10 × restriction buffer (NEBuffer 4, supplied with NlaIII)
- 10 mg/ml BSA (supplied with NlaIII)
- LoTE: 3 mM Tris–HCl pH 7.5, 0.2 mM EDTA pH 7.5

Method

1. If cDNA generated in *Protocol 2* was stored at −20°C, allow it to defrost slowly on ice.

2. Remove the second strand reaction mixture from the tube and rinse once with 50 μl washing solution, then remove washing solution.

3. Rinse tube once with 50 μl of 1 × restriction buffer.

4. Remove buffer, then add:
   - 2.5 μl of 10 × restriction buffer (NEBuffer 4)
   - 0.25 μl BSA (10 mg/ml)
   - 20.25 μl LoTE
   - 2 μl NlaIII (10 U/μl)
   - Final volume 25 μl

5. Digest at 37°C for 1 h.

6. Heat inactivate the restriction enzyme by incubating at 65°C for 20 min.

[a] Several different combinations of anchoring enzymes and tagging enzymes can be used. Here we use the combination of NlaIII as anchoring enzyme and BsmFI as tagging enzyme, as previously described (1).

Protocol 4

Preparation and ligation of linkers

Reagents

- Polynucleotide kinase (PNK) (Pharmacia)
- 10 × PNK buffer (supplied with PNK)
- 10 mM ATP
- LoTE (see *Protocol 3*)
- T4 DNA ligase (5 U/μl; Gibco BRL)
- 5 × T4 DNA ligase buffer (supplied with ligase)
- Primer *linker 1A*:

5′-TTTGGATTTGCTGGTGCAGTACAACTAGG CTTAATAGGGACATG-3′

- Primer *linker 1B*:

5′-TCCCTATTAAGCCTAGTTGTACTGCACCA GCAAATC [amino mod.C7]-3′

- Primer *linker 2A*:

5′-TTTCTGCTCGAATTCAAGCTTCTAACGATG TACGGGGACATG-3′

- Primer *linker 2B*:

5′-TCCCCGTACATCGTTAGAAGCTTGAATTC GAGC [amino mod. C7]-3′

Protocol 4 continued

A. Preparation of linkers

The linkers are constructed by allowing two complementary primers to anneal. First it is necessary to phosphorylate the 5′ end of primers *linker 1B* and *2B*.

1. Dilute all primers to a concentration of 350 ng/μl.

2. Phosphorylate the 5′ end of primers *linker 1B* and *2B* by adding together:

	Tube 1	Tube 2
• Primer *linker 1B* (350 ng/μl)	36 μl	–
• Primer *linker 2B* (350 ng/μl)	–	36 μl
• 10 × PNK buffer	8 μl	8 μl
• 10 mM ATP	4 μl	4 μl
• PNK (9.3 U/μl)	2 μl	2 μl
• LoTE	30 μl	30 μl
• Final volume	80 μl	80 μl

3. Incubate at 37 °C for 30 min.

4. Heat inactivate PNK at 65 °C for 10 min.

5. Add 36 μl of primer *linker 1A* to the 80 μl of kinased primer *linker 1B*. Do the same for primers *linker 2A* and *linker 2B* (final concentration: 217 ng/μl).

6. Anneal primers by heating to 95 °C for 2 min, and subsequently incubating at 65 °C for 10 min, at 37 °C for 10 min, and at room temperature for 20 min.

7. Test linkers by ligating 200 ng of each linker followed by electrophoresis on a 12% polyacrylamide gel. If phosphorylation reaction was successful the majority ($>$ 70%) of the material should have ligated to form linker dimers ($\pm$ 80 bp).

B. Ligation of linkers

1. Remove the reaction mixture from *Nla*III digestion (*Protocol 3*) from the PCR tube and rinse once with 50 μl washing solution (mRNA Capture Kit).

2. Remove washing solution and rinse tube once with 50 μl of 1 × T4 DNA ligase buffer, then remove buffer.

3. Add to the tube (on ice!):
   • 2.5 μl linker 1 ($\pm$ 200 ng/μl)
   • 2.5 μl linker 2 ($\pm$ 200 ng/μl)
   • 5 μl of 5 × T4 DNA ligase buffer
   • 15 μl LoTE
   Final volume 25 μl

4. Anneal linkers to *Nla*III digested cDNA by heating at 50 °C for 2 min, then incubating at room temperature for another 15 min.

5. Add 1 μl of T4 DNA ligase (5 U/μl), then ligate linkers at 16 °C for 2 h.

Protocol 5

Tag release by digestion with tagging enzyme

Reagents

- *Bsm*FI (2 U/μl; New England Biolabs)
- 10 mg/ml BSA (supplied with *Bsm*FI)
- 10 × restriction buffer (NEBuffer 4, supplied with *Bsm*FI)
- LoTE (see *Protocol 3*)
- Phenol:chloroform:isoamyl alcohol (25:24:1)

- 20 mg/ml glycogen (Boehringer Mannheim)
- 10 M ammonium acetate
- Ethanol
- 70% ethanol

Method

1. Remove the ligation reaction mixture from the PCR tube and rinse once with 50 μl washing solution.

2. Remove washing solution and rinse tube once with 50 μl of 1 × restriction buffer, then remove buffer.

3. Add to tube:
   - 2.5 μl of 10 × restriction buffer (NEBuffer 4)
   - 0.25 μl BSA (10 mg/ml)
   - 21.25 μl LoTE
   - 1 μl *Bsm*FI (2 U/μl)
   - Final volume 25 μl

4. Digest at 65 °C for 1 h.

5. Add 175 μl LoTE and transfer to a 1.5 ml Eppendorf tube. Do not discard since mixture contains the released SAGE tag attached to one of either linkers!

6. Add an equal volume phenol:chloroform:isoamyl alcohol.

7. Vortex.

8. Centrifuge in a microcentrifuge at 4 °C for 5 min at 13 000 r.p.m.

9. Transfer aqueous (top) phase to a new 1.5 ml Eppendorf tube.

10. Ethanol precipitate by adding 3 μl glycogen, 100 μl of 10 M ammonium acetate, and 700 μl ethanol.

11. Place at −20 °C for 30 min.

12. Centrifuge in a microcentrifuge at 4 °C for 15 min at 13 000 r.p.m.

13. Wash the pellet twice with 500 μl of 70% ethanol by vigorous vortexing.

14. Remove 70% ethanol and allow pellet to air dry for approx. 15 min.

15. Resuspend pellet in 22.5 μl LoTE.

Protocol 6

Blunting tags

Reagents

- 5 × second strand buffer: 100 mM Tris–HCl pH 6.9, 450 mM KCl, 23 mM $MgCl_2$, 0.75 mM β-NAD$^+$, 50 mM $(NH_4)_2SO_4$ (Gibco BRL)
- dNTPs (25 mM each) (HT Biotechnology)
- Klenow (5 U/μl; Amersham Life Science)
- 10 mg/ml BSA
- LoTE (see *Protocol 3*)

Method

1. Add to the precipitated cDNA tags:
   - 22.5 μl cDNA tags
   - 6 μl of 5 × second strand buffer
   - 0.5 μl BSA (10 mg/ml)
   - 0.5 μl of 25 mM dNTPs
   - 0.5 μl Klenow (5 U/μl)
   - Final volume 30 μl

2. Incubate at 37 °C for 30 min.

3. Raise volume to 200 μl by addition of 170 μl LoTE.

4. Extract with phenol:chloroform:isoamyl alcohol (25:24:1) and ethanol precipitate as described in *Protocol 5*.

5. Resuspend pellet in 8 μl LoTE.

Protocol 7

Ligation to ditags

Reagents

- T4 DNA ligase (5 U/μl; Gibco BRL)
- 5 × T4 DNA ligase buffer (supplied with ligase)
- H_2O

Method

1. Divide the resuspended ditags between two 0.2 ml PCR tubes. One tube serves as a 'no ligase control' to ensure that any amplified ditags in the PCR step are really derived from ligated ditags and not from a contamination.

2. Ligate the blunted tags to ditags by addition of:

Protocol 7 continued

	Sample	No ligase control
• Blunted ditags	4 μl	4 μl
• 5 × T4 DNA ligase buffer	1.2 μl	1.2 μl
• T4 DNA ligase (5 U/μl)	0.8 μl	–
• H₂O	–	0.8 μl
• Final volume	6 μl	6 μl

3. Incubate overnight at 16°C.

Protocol 8
PCR amplification

Equipment and reagents

- Mini-Protean II vertical electrophoresis system (Bio-Rad)
- PCR apparatus
- AmpliTaq (5 U/μl; Perkin Elmer)
- 10 × PCR buffer: 166 mM $(NH_4)_2SO_4$, 670 mM Tris pH 8.8, 67 mM $MgCl_2$, 100 mM 2-mercaptoethanol
- dNTPs (25 mM each) (HT Biotechnology)
- Primer *SAGE 1*:
5'-GGATTTGCTGGTGCAGTACA-3'
- Primer *SAGE 2*:
5'-CTGCTCGAATTCAAGCTTCT-3'
- H₂O
- DMSO (Sigma)
- Phenol:chloroform:isoamyl alcohol (25:24:1)

- 20 mg/ml glycogen (Boehringer Mannheim)
- 10 M ammonium acetate
- Ethanol
- 70% ethanol
- 40% polyacrylamide solution: 19:1 acrylamide:bisacrylamide (Bio-Rad)
- TEMED
- 10% ammonium persulfate
- 10 bp ladder (Gibco BRL)
- 5 × loading dye: 20% sucrose, 0.5% SDS, 50 mM EDTA, 50 mM Tris–HCl pH 7.4, 0.125% bromophenol blue
- 50 × TAE

Method

1. Raise the volume of the ligation mixture to 20 μl by adding LoTE.

2. Prepare a 1/100 dilution of the ligation mixture.

3. Use 1 μl of the 1/100 dilution as template in a 50 μl PCR reaction. Perform seven parallel identical 50 μl PCR reactions:
   - 5 μl of 10 × PCR buffer
   - 2 μl dNTPs (25 mM each)
   - 3 μl DMSO
   - 1 μl primer *SAGE1* (350 ng/μl)

Protocol 8 continued

- 1 µl primer *SAGE2* (350 ng/µl)
- 36 µl ddH$_2$O
- 1 µl Ampli*Taq* (5 U/µl)
- Final volume 49 µl

4. Finally add 1 µl of the diluted ligation mix as template in the PCR reaction.

5. Also dilute the 'no ligase control' 1/100 and use 1 µl in a 50 µl PCR reaction. Perform PCR in duplo, amplify one reaction for 28 cycles and the other for 35 cycles.

6. PCR amplify using the following conditions:

   (a) One cycle of 5 min at 95 °C.

   (b) 28 cycles of 30 sec at 95 °C, 1 min at 55 °C, 1 min at 70 °C.

   (c) One cycle of 5 min at 70 °C.

   (d) Room temperature.

7. After PCR amplification, pool the seven identical PCR reactions and extract with an equal volume of phenol:chloroform:isoamyl alcohol.

8. Ethanol precipitate as described in *Protocol 5* using 3 µl glycogen, 100 µl of 10 M ammonium acetate, and 1 ml ethanol. Resuspend pellet in 100 µl LoTE.

9. Add 20 µl of 5 × loading dye.

10. Divide between five lanes of a 12% polyacrylamide gel in 1 × TAE. Use a 10 bp ladder as marker. For electrophoresis we recommend the Mini-Protean II vertical electrophoresis system (Bio-Rad).

There is no easy way to control that the steps leading to the formation of ditags have worked. The first visible evidence is after PCR amplification of the ditag, when a product of ± 100 bp should be visible, which represents the ditag (22–26 bp) with a linker at both ends (2 × ± 40 bp). A background product of ± 80 bp is always visible with an almost equal intensity to the ditag product, and sometimes also a 90 bp product consisting of a single tag (monotag) joined to both linkers. If the amount of starting material is limited (less than 1 µg of poly(A)$^+$ RNA), probably there will be hardly any 100 bp product visible after 28 cycles of PCR. In this case the region around 100 bp is excised from the gel, the DNA is extracted and precipitated, and subjected to a limited number of re-PCR cycles. It is important to empirically determine the optimal number of re-PCR cycles by performing a cycle series. The number of cycles of re-PCR should, however, be restricted as much as possible to avoid preferential amplification of certain ditag species.

In SAGE it is of the utmost importance to prevent PCR contamination, since every experiment yields PCR products of exactly the same length. 'No template controls' are extremely important in microSAGE, due to the large number of PCR cycles that are performed. We also recommend a strict separation of pre-

PCR and post-PCR sample handling, the use of filter tips to prevent contamination with ditags generated in previous experiments, and aliquoting of all PCR reaction components. Due to inter-PCR machine variation it is advisable to optimize the PCR conditions for use in your own PCR machine. If the used PCR apparatus does not have a heated lid, add mineral oil on top of the reaction mixture to prevent evaporation.

Protocol 9
Re-PCR

Equipment and reagents

- See *Protocol 8*
- Scalpel or razor blade
- 21 gauge needle
- LoTE (see *Protocol 3*)
- SPIN-X columns, 0.22 μm (CoStar)
- 10 M ammonium acetate

- Phenol:chloroform:isoamyl alcohol (25:24:1)
- 20 mg/ml glycogen (Boehringer Mannheim)
- Ethanol
- 70% ethanol

Method

1. Excise region containing ditag (region around 100 bp) across all five lanes of the gel using a clean scalpel or razor.[a]

2. Divide the excised polyacrylamide fragments between two 0.5 ml PCR tubes pierced with a 21 gauge needle, and place the PCR tubes in a 1.5 ml Eppendorf tube. Centrifuge for 5 min at 13 000 r.p.m. to fragment the polyacrylamide.[b]

3. Remove the 0.5 ml tube and add 300 μl LoTE to both 1.5 ml tubes with the polyacrylamide fragments.

4. Incubate for 20 min at 65 °C.

5. Transfer contents of tube to a SPIN-X column and spin in a microcentrifuge at 13 000 r.p.m. for 5 min.

6. Transfer the filtrates to a new 1.5 ml Eppendorf tube and ethanol precipitate as described in *Protocol 8*.

7. Resuspend both pellets in a total volume of 1 ml LoTE.

8. Use 1 μl as template in a 50 μl re-PCR with primers *SAGE1* and *SAGE2* as described in *Protocol 8*. Perform a cycle series to determine the optimal number of re-PCR cycles (usually between 8–15) (*Figure 4*).

[a] To avoid contamination between SAGE libraries, remove any trace of ditags of previous experiments by soaking the electrophoresis chamber, glass plates, spacers, combs, etc. in 0.5 M NaOH for at least 30 min before running the PCR-amplified ditags on gel.

[b] Do not overfill the pierced PCR tubes with excised polyacrylamide slices. Use a new tube if excised fragment exceeds three lanes of the gel in width.

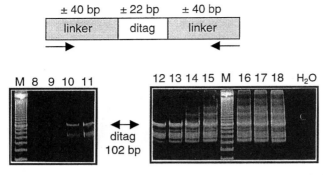

Figure 4 Ditag PCR products run on a 12% polyacrylamide gel. Optimizing the number of PCR cycles is important in the SAGE procedure by increasing the number of PCR cycles by one at a time. The ditag band of approximately 100 bp is clearly visible, as well as the most prominent background product of around 80 bp. At 12 cycles in this example the ratio of ditag to background product is optimal. Too many PCR cycles can result in a decrease of the amount of visible ditag due to exhaustion of the PCR reaction and consequent denaturation of formed products. M: marker (100 bp ladder); this marker has a band every 10 bp with the most intense marker band as 100 bp.

Protocol 10

Large scale amplification of ditag

Equipment and reagents

- See *Protocol 8*
- Ethidium bromide

Method

1. Once the optimal number of cycles of re-PCR has been determined, 96 parallel 100 μl PCRs are performed using the same conditions, only scaled up from 50 μl to 100 μl.

2. Pool the 100 μl PCRs per eight tubes in a 2 ml Eppendorf tube.

3. Extract with an equal volume of phenol:chloroform:isoamyl alcohol (25:24:1).

4. Ethanol precipitate by adding 3 μl glycogen, 100 μl of 10 M ammonium acetate, and 1 ml ethanol per 300 μl PCR mixture.

5. Resuspend the pellets (approx. 30 identical tubes!) sequentially in a total volume of 250 μl LoTE. Add 50 μl loading dye.

6. Divide over 18 lanes of two 12% polyacrylamide gels (ten lanes per gel).

7. After electrophoresis, stain gels with ethidium bromide (*Figure 5*).[a]

8. Excise ditag band from the gels (approx. 100 bp) and purify as described in *Protocol 9*, only using a total of six SPIN-X columns.

9. After ethanol precipitation, resuspend the pellets in a total volume of 300 μl LoTE.

[a] Stain the gel containing the ditags generated by large scale PCR amplification with the less sensitive ethidium bromide instead of SYBR Gold.

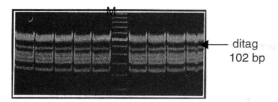

— ditag
102 bp

Figure 5 Preparative gel with large scale re-PCR of ditag. The ditag of ~ 100 bp is excised and the DNA is purified from the gel fragment.

After large scale PCR amplification and purification of the100 bp product the linkers are cleaved off to form the small 22–26 bp ditag. Due to its short length this product is easily denatured, which is detrimental because reassociation does not occur easily in the extremely heterogeneous population of molecules. It is therefore important to take precautions to prevent denaturation of the ditag, by performing all centrifugation steps in a chilled microcentrifuge and not heating it above 37 °C.

If you encounter difficulties cleaving off the linkers, there might be a problem with NlaIII. NlaIII is fairly unstable at −20 °C; therefore it is recommended to aliquot the enzyme and store at −80 °C; only keep a small quantity at −20 °C.

Protocol 11

Cleaving off the linkers by digestion with the anchoring enzyme

Reagents

- See *Protocols* 3 and 5

Method

1. Divide the 300 μl of ditags between two tubes of 150 μl each.

2. To each tube add:
   - Ditags .... 150 μl
   - 10 × restriction buffer (NEBuffer 4) ..... 20 μl
   - BSA (10 mg/ml) .... 4 μl
   - NlaIII (10 U/μl) .... 10 μl
   - LoTE .... 16 μl
   - Final volume .... 200 μl

3. Digest at 37 °C for 2 h.

4. Extract with an equal volume of phenol:chloroform:isoamyl alcohol (25:24:1).

5. After addition of 3 μl glycogen, 100 μl of 10 M ammonium acetate, and 700 μl ethanol per tube, chill for 15 min in a dry ice/ethanol bath or overnight at −80 °C before centrifugation.

Protocol 11 continued

6. Resuspend both pellets in a total volume of 30 μl LoTE.

7. Load the sample on two lanes of a 12% polyacrylamide gel.[a,b] Use a 10 bp ladder as a marker. We recommend using a loading dye containing Orange G as dye to prevent co-migration with the ditag of 22–26 bp.

8. Run the gel at room temperature at 75 V for 5 h or until the Orange G dye is at the bottom of the gel.

9. Stain gel with SYBR Gold (Molecular Probes).

10. Excise the ditag band from gel. This band runs at 22–26 bp (*Figure 6*) just below the linkers that run at ± 40 bp.

11. Purify the ditag from the gel slice as described in *Protocol 9*, only incubate the polyacrylamide fragments for 20 min at 37 °C instead of 65 °C.[c] Proceed immediately to the ethanol precipitation step without a phenol:chloroform:isoamyl alcohol extraction.

12. Add 3 μl glycogen, 100 μl of 10 M ammonium acetate, and 1 ml ethanol. Store at −80 °C overnight before the centrifugation step.

13. Resuspend the pellet(s) in a total volume of 7.5 μl LoTE.[d]

[a] It is advisable to use longer acrylamide gels to get a better separation of the excised ditag from the linkers. Make sure lanes on the gel are not overloaded and that the ditag band of 22–26 bp runs separately from the linker band at 40 bp on the gel. The presence of trace amounts of linkers can poison the concatenation and prevent cloning.

[b] Be careful not to introduce any traces of DNase in the sample. The presence of even a small proportion of ditags with only one intact sticky end can be detrimental for the cloning efficiency. These 'faulty' ditags can be added to the concatemers, but do not have another intact sticky end left to add on another ditag. This not only results in termination of concatenation, but also prevents ligation in the cloning vector.

[c] The lower incubation temperature is important to prevent denaturation of the small 22–26 bp ditag. Once denatured, the ditags will remain single stranded due to the enormous heterogeneity of the ditag population.

[d] It may be necessary to repeat *Protocols 10* and *11* to generate sufficient ditag for concatenation. Generally we perform 200–300 100 μl PCRs in microSAGE.

Protocol 12

Concatenation of ditags and size selection

Equipment and reagents

- See *Protocols 4* and *8*
- 100 bp marker

Method

1. Ligate the resuspended ditags to concatemers by addition of ligase:
   - 7.5 μl ditags
   - 2 μl of 5 × T4 DNA ligase buffer

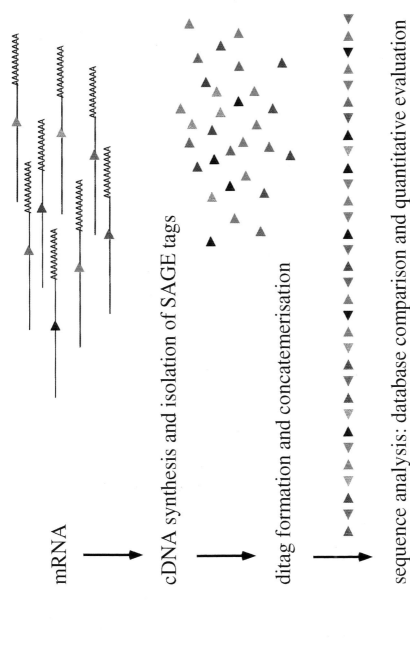

mRNA

cDNA synthesis and isolation of SAGE tags

ditag formation and concatemerisation

sequence analysis: database comparison and quantitative evaluation

Chapter 9: Fig. 1 Simplified scheme of SAGE. Short ~10 bp sequence tags are isolated from cDNA, thus maintaining the original stoichiometry of the individual cDNA molecules. The tags are subsequently ligated to long tag-multimers, which are cloned and sequenced. The frequency of each tag in the cloned multi-mers is a quantitative measure of gene expression. In addition, the sequence of each tag is transcript-specific, facilitating identification of the corresponding gene.

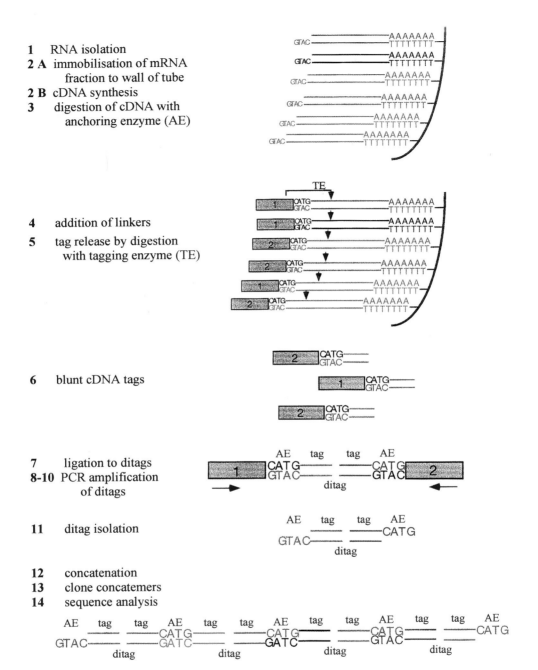

1	RNA isolation
2 A	immobilisation of mRNA fraction to wall of tube
2 B	cDNA synthesis
3	digestion of cDNA with anchoring enzyme (AE)
4	addition of linkers
5	tag release by digestion with tagging enzyme (TE)
6	blunt cDNA tags
7	ligation to ditags
8-10	PCR amplification of ditags
11	ditag isolation
12	concatenation
13	clone concatemers
14	sequence analysis

Chapter 9: Fig. 2 Flow chart of the SAGE procedure, outlining all the different steps involved (1 to 14) (see text). Each step in the procedure is detailed in its corresponding protocol (*Protocols 1–15*).

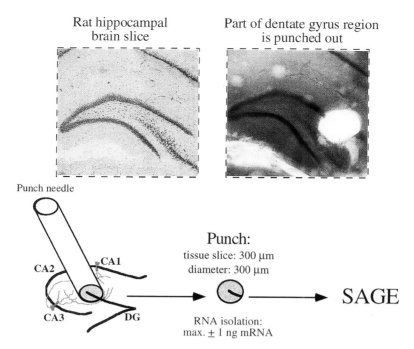

Rat hippocampal
brain slice

Part of dentate gyrus region
is punched out

Punch needle

CA2 | CA1

CA3 | DG

Punch:
tissue slice: 300 μm
diameter: 300 μm

RNA isolation:
max. ± 1 ng mRNA

SAGE

Chapter 9: Fig. 3 Example of punching of the rat dentate gyrus. The 75μm section (*left*) is localized immediately adjacent to the 300μm section (*right*). This 75μm section was stained with cresyl violet to visualize
the cellular layers, thus facilitating punching out of the correct region, in this case the tip of the inner layer
of the rat dentate gyrus. After punching, the 300μm section was also stained to visualize the punched area.
Correctly punched out regions are used as input in the microSAGE procedure. The lower part of the figure is
a schematic description of the punch procedure (see also ref. 7).

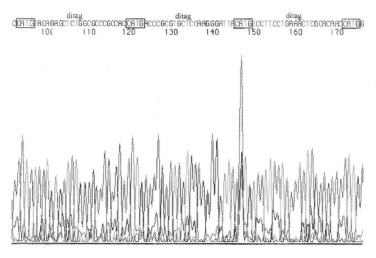

Chapter 9: Fig. 9 Example of a part of a sequence of a cloned concatemer, showing ditags separated by
the CATG-sequence which is the recognition site for the anchoring enzyme *Nla*III.

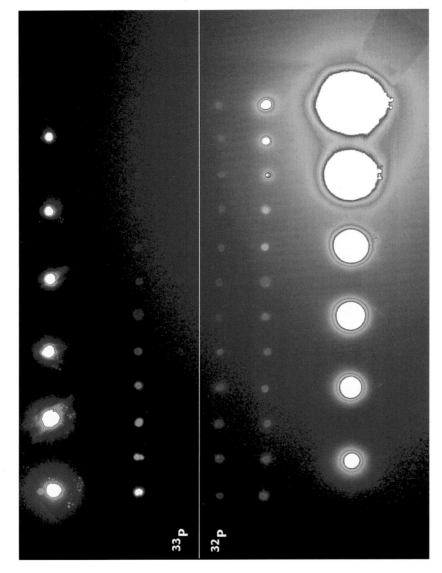

Chapter 10: Fig. 2 Signal dispersion with ^{32}P versus ^{33}P. Identical threefold serial dilutions of dCT^{32}P and dCT^{33}P were made and spotted onto a nylon membrane. The first spot of each series is 0.66 μCi; each subsequent spot contains one-third the amount of material in the previous spot. The membrane was exposed to a PhosphorImaging screen for 24 h and then scanned with a Storm PhosphorImager (Molecular Dynamics). The actual distance between the spots in the bottom and top rows is approximately 18 mm centre-to-centre. The spots in the remaining rows are approximately 9 mm centre-to-centre. Note that the first spot on the ^{33}P series is on the top left with the serial dilutions proceeding from left to right, down and again left to right. The ^{32}P series begins on the bottom right and the serial dilutions go from right to left, up and again right to left.

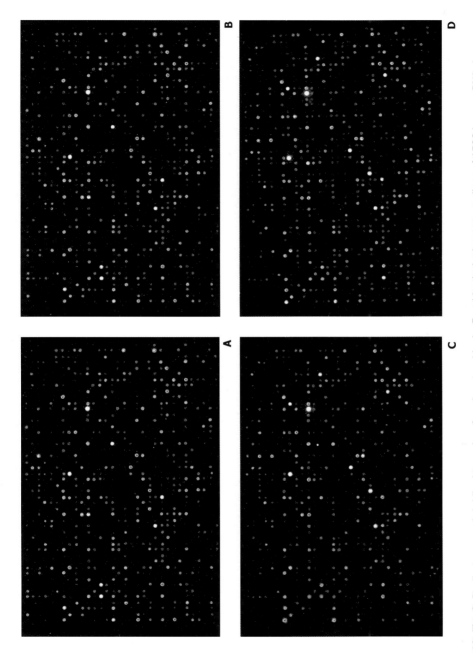

Chapter 10: Fig. 3 Duplicate arrays hybridized to probes from heart and brain. Four identical microarray replicas of 1536 human cDNA clones were prepared on nylon membranes as described in *Protocol 2*. RNA was purchased from Clontech (brain, Cat. No. 6516-1; or heart, Cat. No. 6533-1). RNA was labelled using dCT^{33}P as described in *Protocol 5*. Hybridization was performed in a sealed box for 16 h as described in *Protocol 6*. After washing, the membranes were exposed to a PhosphorImaging screen and then scanned with a Storm PhosphorImager (Molecular Dynamics). The image shown is false-coloured to better show different levels of signals from low to high; ranging from black (lowest) to blue, green, yellow, orange, and red (highest). (A and B) Duplicate hybridizations to total heart RNA. (C and D) Duplicate hybridizations to total brain RNA.

Chapter 10: Fig. 4 *ArrayVision* grid placement. The PhosphorImage from *Figure 3A* was analysed using the *ArrayVision* software (Imaging Research). A grid was specified and then placed by the program to overlay all spots in the microarray. The labels and grid show that 16 96-well plates are arrayed on this membrane.

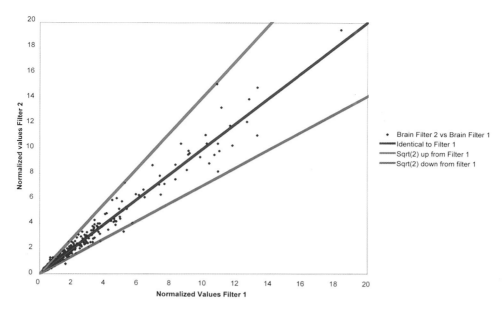

Chapter 10: Fig. 5 Scatter plot of duplicates. The *ArrayVision* software was used to quantitate the arrays shown in *Figures 3C* and *3D*. The values were normalized as described in Section 5.2.3. The blue line indicates where identical values would fall. The area between the green and red lines represents twofold variation.

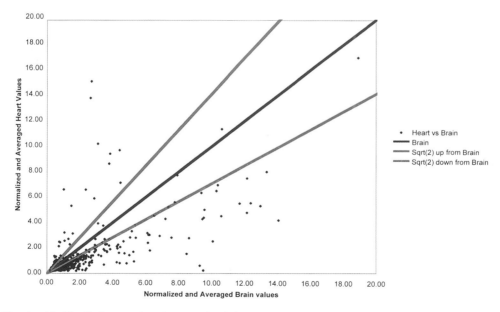

Chapter 10: Fig. 7 Scatter plot of expression in heart versus brain. The *ArrayVision* software was used to quantitate all arrays in *Figure 3*. All values were normalized as described in Section 5.2.3 and then duplicate values were averaged. Values above the green line correspond to genes more highly expressed in heart than brain, while values below the red line correspond to genes more highly expressed in brain than heart.

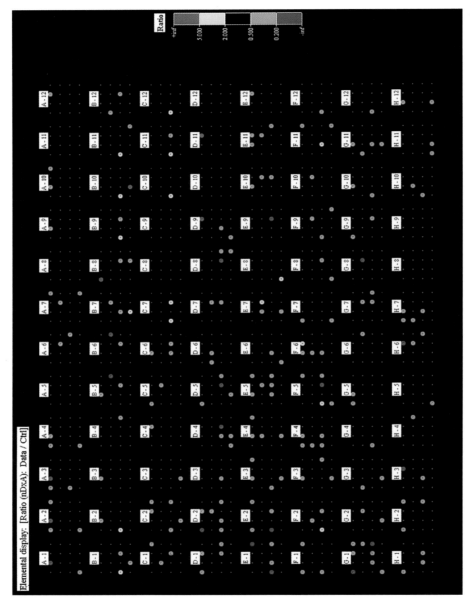

Chapter 10: Fig. 8 Elemental display of expression in heart versus brain. The *ArrayVision* software was used to quantitate and normalize the arrays in *Figures 3A* and *3C*. Normalized values were compared using the elemental display feature of *ArrayVision* with brain defined as 'control' and heart defined as 'data'. Values are compared at each element on the array. Spots in black show less than twofold change up or down between the brain and heart samples. Spots in yellow are higher in heart by two- to fivefold. Red elements indicate a greater than fivefold. Light and dark blue elements indicate a two- to five- and greater than fivefold increase in brain, respectively.

Protocol 12 continued

- 0.5 μl T4 DNA ligase (5 U/μl)
- Final volume 10 μl

2. Ligate for 30 min at 16 °C.

3. Load the concatenated ditags in a single lane of a 8% polyacrylamide gel. Use a 100 bp ladder as marker. Run the gel at room temperature at 120 V for 5 h.[a]

4. Stain the gel with SYBR Gold (*Figure 7*).

5. Excise the regions of the gel which contain concatemers of 200–300 bp, 300–400 bp, 400–800 bp, and larger than 800 bp.

6. Purify DNA from the excised gel fragments as described in *Protocol 9*.[b]

7. After precipitation resuspend the pellets each in 5 μl LoTE.

[a] Ligation of ditags to concatemers should give a smear on a 8% polyacrylamide gel ranging from the bottom of the gel to well above 1 kb. If this is not the case some linkers might have been accidentally isolated along with the ditag during the gel separation, giving rise to premature termination of the concatemers.

[b] In the first instance the 400–800 bp fraction is used for cloning. Keep the other excised fractions stored under ethanol (do not centrifuge until required). These fractions are back-ups which can be used if the cloning of the 400–800 bp fraction fails or does not yield sufficient colonies.

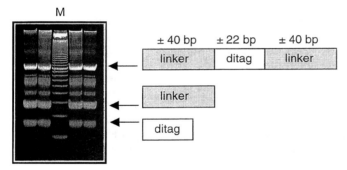

Figure 6 A 12% polyacrylamide gel containing cleaved ditag. The ditag band of 22–26 bp is visible after staining of the gel. Also visible are the linkers of approximately 40 bp, migrating just above the ditag, and remnants of undigested ditag around 100 bp. In addition, there are some other background products which may result from partially digested ditag.

Protocol 13

Cloning of size-selected concatemers

Equipment and reagents

- pZero Background™ Cloning Kit (Invitrogen)
- Electroporation cuvettes (BTX; 1 mm gap)
- Gene Pulser II System (Bio-Rad)
- 37 °C incubator
- *Sph*I (10 U/μl; Promega)
- ELECTROMAX DH10B™ Cells (Gibco BRL)
- SOC medium
- LB agar plates with Zeocin

Protocol 13 continued

A. Ligation of concatemers in the SphI site of pZero

1. Ligate the size-selected concatemers in a vector of choice digested with *Sph*I. We recommend cloning of concatemers in the *Sph*I site of pZero. For specific details on cloning in this vector we refer you to the manual of the Zero Background Cloning Kit.

   - 5 μl size-selected concatemers in LoTE
   - 1 μl pZero × *Sph*I (20 ng/μl)
   - 1 μl of 10 × T4 DNA ligase buffer (pZero Background Cloning Kit)
   - 0.5 μl T4 DNA ligase (4 U/μl) (pZero Background Cloning Kit)
   - 2.5 μl H₂O (pZero Background Cloning Kit)
   - Final volume 10 μl

2. Ligate at 16 °C for 30 min.

3. Add 10 μl H₂O (pZero Background Cloning Kit).

4. Heat inactivate ligase by incubating for 5 min at 65 °C.

5. Use 2 μl of ligation to transform ELECTROMAX DH10B™ Cells.[a]

B. Transformation of E. coli[b]

1. Allow a vial of 100 μl ELECTROMAX DH10B™ Cells to thaw on ice.

2. Meanwhile chill the required number of electroporation cuvettes on ice.[c]

3. When thawed, immediately aliquot the ELECTROMAX DH10B™ Cells into 35 μl portions (keep on ice!).

4. Add 2 μl of ligation mix to a 35 μl portion of cells; mix by gently tapping the tube.

5. Transfer the cells to the chilled cuvette (keep on ice!).

6. Electroporate in a Gene Pulser II using the settings: 2.5 kV, 100 Ohm, 25 μF.

7. Immediately after applying the pulse, add 465 μl of SOC medium.

8. Incubate at 37 °C for 1 h.

9. Plate out in 100 μl portions on LB agar plates containing the antibiotic Zeocin.[d]

10. Incubate the plates overnight in a 37 °C incubator.

11. The next day the plates should contain hundreds of colonies.

[a] It is advisable to perform preparation of the vector, ligation, and transformation on the same day.

[b] Any transformation procedure can be used, but we recommend electroporation due to its high transformation efficiency in combination with the extremely competent ELECTROMAX DH10B™ Cells (> 10^{10} transformants/μg plasmid).

[c] We generally perform multiple parallel electroporations, each with 2 μl of ligation mix to generate a SAGE library with sufficient tag potential.

[d] The details for the plates and SOC medium are described in the manual of the Zero Background Cloning Kit.

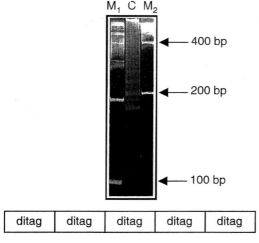

Figure 7 Concatenated ditags run on an 8% polyacrylamide gel, showing a smear ranging from below 100 bp to over 1 kb. The markers are the 10 bp ladder, with bands every 10 bp up to 330 bp and a large fragment high in the gel at 1668 bp, and the 100 bp ladder with a band every 100 bp up to 1200 bp.

Protocol 14

Sequencing of cloned concatemers

Equipment and reagents

- PCR apparatus
- BigDye Terminator Cycle Sequencing Ready Reaction Kit (Perkin Elmer)
- ABI 377 automated sequencer (Perkin Elmer)
- Super*Taq* and 10 × Super*Taq* buffer (HT Biotechnology)
- dNTPs (HT Biotechnology)
- *EV2*:
  5′-CGCCAGGGTTTTCCCAGTCACGAC-3′

- *EV3*:
  5′-AGCGGATAACAATTTCACACAGGA-3′
- *SP6*:
  5′-TATTTAGGTGACACTATAG-3′
- 10 × sequencing dilution buffer: 200 mM Tris pH 9.0, 5 mM $MgCl_2$
- 95% ethanol
- Formamide loading dye

A. Insert PCR

1. Pick individual colonies into 100 µl H_2O.
2. Use 1 µl of the bacterial suspension in a PCR with primers *EV2* and *EV3*:
   - 2.5 µl of 10 × Super*Taq* buffer
   - 0.2 µl dNTPs (25 mM each)
   - 1 µl primer *EV2* (50 ng/µl)
   - 1 µl primer *EV3* (50 ng/µl)

Protocol 14 continued

- 19.1 μl H$_2$O
- 1 μl bacterial suspension
- 0.2 μl Super*Taq* (5 U/μl)
- Final volume 25 μl

3. PCR amplify using the following conditions:
    (a) One cycle of 5 min at 94 °C.
    (b) 28 cycles of 30 sec at 94 °C, 30 sec at 61 °C, 1.5 min at 72 °C.
    (c) One cycle of 5 min at 72 °C.
    (d) 4 °C.

4. Analyse PCR products on a 1% agarose gel (*Figure 8*). The minimal length of most PCR products should be at least 500 bp (this includes approx. 270 bp of vector sequence) in order for them to contain an average of at least 15 tags. Smaller PCR products will yield fewer tags per sequence reaction.

5. Once sufficient clones with inserts have been obtained it is advisable to store individual colonies in separate wells of 96-well microtitre dishes filled with LB + 7% glycerol. After O/N incubation at 37 °C, the microtitre dishes can be stored at −80 °C until required for sequence analysis.

6. For sequencing, allow glycerol stock to thaw, then dilute 2 μl by adding 50 μl H$_2$O. Use 1 μl of dilution as template in insert PCR.

B. Sequence reactions

1. Direct sequencing of PCR products is performed without any further purification step.

2. By diluting the ready reaction mix with the dilution buffer, the costs of sequencing can be reduced significantly. In our experience a fourfold dilution works fine on unpurified PCR products as template in the sequencing reaction, yielding a fourfold increase in amount of sequence reactions in a BigDye Terminator Kit.

3. Use 1 μl of PCR product directly as template for cycle sequencing:
    - 2 μl Terminator Ready Reaction Mix
    - 6 μl of 10 × sequencing dilution buffer
    - 0.6 μl SP6 (10 μM)
    - 10.4 μl H$_2$O
    - 1 μl PCR product (20–50 ng)
    - Final volume 20 μl

4. Use the following PCR conditions:
    (a) 26 cycles of 30 sec at 96 °C, 5 sec at 50 °C, 4 min at 60 °C.
    (b) 4 °C.

5. Ethanol precipitate the sequence reactions by adding 16 μl H$_2$O and 64 μl of 95% ethanol. Incubate for 30 min at room temperature and then centrifuge for 20 min at 13 000 r.p.m. at 4 °C.

Protocol 14 continued

6. Resuspend pellets in 6 μl of formamide loading dye.

7. Denature sequence products at 95 °C for 3 min and place directly on ice.

8. Load 1.5 μl on a 5% Long Ranger sequence gel.

9. The raw sequence data is analysed using the SAGE program group software (please see *Figure 9* in plate section) (1, 5).

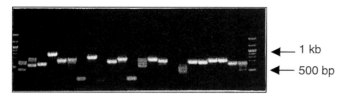

← 1 kb

← 500 bp

Figure 8 An example of insert PCR on colonies with cloned concatemers. Most PCR products are larger than 500 bp, and should contain a minimum of 15 tags. A few colonies only yield a PCR product consisting of vector sequences (~ 270 bp) and thus do not have an insert.

Protocol 15

Analysis of sequence data

Equipment

• SAGE program group software[a]

Method

1. The raw sequence data can be analysed using the SAGE program group software, which locates the anchoring enzyme sites in the sequence data, extracts ditags between two adjacent anchoring enzyme sites, and then extracts individual tags from the ditags.

2. Each ditag included in the analysis is a unique combination of two individual tags, since the probability of formation of the same ditag is extremely low. This ensures that any (PCR-based) bias resulting in preferential formation or amplification of specific ditag species does not interfere with the obtained SAGE profile.

3. The profile can be linked to GenBank data to determine the identity of the tags. Comparison of different profiles will result in the identification of differentially expressed tags.

4. An example of a comparison of two partial microSAGE profiles derived from dentate gyrus punches is shown in *Figure 10*.

[a] The software is available from Dr Kinzler of the Johns Hopkins Oncology Center, Molecular Genetics Laboratory, Baltimore, USA.

Example of microSAGE data: comparison with rat sequences in GenBank

library 1: 1 punch, no apoptosis (1792 tags)
library 2: 6 punches, apoptosis (1606 tags)

tag	Gene (accession no.)	library:	frequency 1	2	percent 1	2
CCTGTAATCC	IFNgamma-induced astrocyte EST (H39351)		12	17	0.67%	1.1 %
CACCTAATTG	unknown		15	1	0.84%	0.06%
GTGAAACCCC	novel G protein-coupled P2 receptor (D63665)		10	3	0.56%	0.19%
TCCTTCCTGA	unknown		10	1	0.56%	0.06%
GACAAGGTCG	growth hormone (E00002)		7	2	0.39%	0.12%
TCTCCATACC	unknown		9	1	0.50%	0.06%
TACCACCTTT	myosin light chain (M12022)		7	4	0.39%	0.25%
AGGTCAGGAG	unknown		5	6	0.28%	0.37%
CTTCTACTAA	unknown		1	9	0.05%	0.56%
AAGGAGATGG	ribosomal protein L31 (X04809)		3	2	0.17%	0.12%
CACAGAGTCC	EST (AA817914)		2	2	0.11%	0.12%
GAAAAAAAAA	EST (AA859996)		2	1	0.11%	0.06%
AATCAACCCG	mitochondrial genes for 16S rRNA and tRNAs (V00681)		1	2	0.05%	0.12%
CCAGAGGCTG	EST (AA899981)		1	2	0.05%	0.12%
GATGCCCCCC	cDNA similar to cytocrome oxidase I (C06520)		1	3	0.05%	0.19%
GCCTCCAAGG	glyceraldehyde-3-phosphate dehydrogenase (M29341)		1	1	0.05%	0.06%
GGGCTGGGGT	EST (AA875618)		2	1	0.11%	0.06%
CCTGTAATCT	unknown		1	3	0.05%	0.19%
TAAAAAAAAA	trachae (Gang Li) cDNA clone 7A3 (U82289) *		1	1	0.05%	0.06%
TGTTATTTAA	cell surface protein (X01785)		1	1	0.05%	0.06%
TTCTCTACCA	EST (AA819729)		1	1	0.05%	0.06%
GTGGCTCACA	glutathione s-transferase (M12894)		0	8	0 %	0.5 %
ATACTGACAC	cytochrome c oxidase (C06609)		0	4	0 %	0.25%
CAGATCTTCG	polyubiquitin (D16554)		0	1	0 %	0.06%
GAAGCAGGAC	cofilin (X62908)		3	0	0.17%	0 %
AATCGCTTCT	transcription factor Stat5b (X97541)		1	0	0.05%	0 %

Figure 10 An example of quantitative and qualitative data obtained by comparing SAGE profiles from dentate gyrus punches with and without apoptotic neurons.

References

1. Velculescu, V. E., Zhang, L., Vogelstein, B., and Kinzler, K. W. (1995). *Science*, **270**, 484.

2. Madden, S. L., Galella, E. A., Zhu, J., Bertelsen, A. H., and Beaudry, G. A. (1997). *Oncogene*, **15**, 1079.

3. Polyak, K., Xia, Y., Zweier, J. L., Kinzler, K. W., and Vogelstein, B. (1997). *Nature*, **389**, 300.

4. Zhang, L., Zhou, W., Velculescu, V. E., Kern, S. E., Hruban, R. H., Hamilton, S. R., *et al.* (1997). *Science*, **276**, 1268.

5. Velculescu, V. E., Zhang, L., Zhou, W., Vogelstein, J., Basrai, M. A., Bassett, D. E., *et al.* (1997). *Cell*, **88**, 243.

6. Palkovits, M. (1973). *Brain Res.*, **59**, 449.

7. Datson, N. A., van der Perk-de Jong, J., van den Berg, M. P., de Kloet, E. R., and Vreugdenhil, E. (1999). *Nucleic Acids Res.*, **27**, 1300.

Chapter 10

Construction and applications of gene microarrays on nylon membranes

JEFFREY L. MOONEY*, PAUL S. KAYNE[†], SHAWN P. O'BRIEN*, and CHRISTINE M. DEBOUCK*

*SmithKline Beecham Pharmaceuticals, 709 Swedeland Road, King of Prussia PA 19406, U.S.A.

[†]Bristol-Myers Squibb, P.O. Box 4500, Princeton, NJ, 08543, U.S.A.

1 Introduction

The rapid advancement of expressed sequence tag (EST) and genomic sequencing has been paralleled by the need to develop methodologies for the functional characterization of newly identified genes. Since the expression pattern of a gene provides direct or indirect information about its function, a number of new approaches have been developed to determine the normal tissue distribution of large numbers of genes and/or to identify genes that are differentially expressed between two or more given tissues or cell types. Some of these methods are designed for the *de novo* isolation of genes expressed at different levels in compared tissues or cell types, e.g. differential display (DD) (1), and related methods such as RNA fingerprinting by arbitrary primer PCR (RAP-PCR) (2), also representation difference analysis (RDA) (3), and suppressive subtractive hybridization (SSH) (4). The other methods exploit high-throughput sequencing and sequence databases either to estimate transcript abundance, e.g. EST numerical analysis (5) and serial analysis of gene expression (SAGE) (6), or to construct DNA microarrays of thousands of gene sequences derived from the databases, e.g. oligonucleotide arrays (7) or gene fragment arrays (8). Individual laboratories need to balance the pros and cons of each technique, including labour-intensiveness and the need for expensive equipment, and decide which of these methodologies is best suited to study their particular biological problem.

The differential display (DD-RTPCR) and related strategies have been applied by many groups to identify small numbers of differentially expressed genes (see ref. 9 for a recent review). These procedures generate large numbers of DNA fragments potentially representing differentially expressed genes but these must be first individually isolated and confirmed (e.g. by Northern blot) prior to further study to avoid working on false positives. Moreover, since the DNA fragments typically range from 110–450 bp containing mainly the 3′ untranslated region (3'UTRs), most of them must be extended into larger, ideally full-length cDNA clones to obtain more useful sequence information. These disadvantages limit the throughput of this type of approach due to the labour-intensive nature of the validation and library screening processes required to identify truly differentially expressed hits with sufficient coding information to determine the identity of the gene. The similar RDA and SSH methodologies have also been used for the detection of differentially expressed genes (3, 4, 10–13). In these methods, both subtractive hybridization and PCR amplification are combined to identify sequences which are greatly reduced or absent in one of the two types of tissues or cells under comparison. While these methodologies are more high-throughput and generate less false positives than DD-RTPCR, both RDA and SSH lead to the identification of short cDNA fragments that need to be subcloned or extended to yield more useful sequence information.

Large scale sequencing and EST analysis has been applied to estimate the tissue distribution of novel genes as well as to identify differentially expressed genes between given cDNA libraries (5). While computation of relative abundancies of gene sequences in a given cDNA library compared to other libraries is now both quick and routine, the quality of such *in silico* analyses is directly dependent upon the depth of sequencing within each of the various libraries compared as well as the overall quality of the libraries themselves. Additionally, the wet work involved in validating these *in silico* analyses can be extremely time-consuming. The SAGE methodology couples high-throughput sequencing to a specialized library construction method (6). In this method, nine base pair gene tags are captured from each mRNA species present in a given tissue sample, linked in long concatemers, and subjected to high-throughput sequencing. A few hundred to a thousand sequencing reactions can identify tags for 10 000 to 50 000 expressed genes from a given RNA source and comparisons of differentially expressed genes between tissue types can be conducted very quickly. However, the true power of this procedure depends on the quality and breadth of the sequence databases used for comparative analysis since only nine base pairs of sequence are available for each gene. In addition, since high-throughput sequencing and sophisticated bioinformatics capabilities are required, this approach is not readily applicable in the majority of academic laboratories.

Array-based technologies allow for the rapid and simultaneous analysis of the expression of very large numbers of genes across a wide variety of tissues, cell types, and conditions whether using arrays of oligonucleotides (7) or gene fragments (8). The technique is amenable to the analysis of a few hundred well defined genes or family members (e.g. all putative kinases or G protein-coupled receptors) as well as to the monitoring of 'global' expression profiles of the entire complement of an organism's expressed genes (14–16). For oligonucleotide microarrays, the numbers of genes studied is dependent upon the number of genes identified through sequencing or *de novo* discovery efforts. Microarrays of gene fragments are typically conducted using genes with available sequence, but one can also readily array inserts from clones randomly picked from a cDNA library (17). In this case, hybridization patterns of interest reveal those gene inserts worthy of sequencing.

This chapter provides methodologies and considerations for the construction and applications of microarrays on nylon membranes. We chose to describe this format since it is currently readily available to the academic and industrial laboratories either through *de novo* construction or through purchase of commercially available products (e.g. Atlas blots by Clontech, human and mouse GEM Arrays by Genomes Systems, GeneFilters™ by Research Genetics). We describe here in detail the procedures that we have adopted, developed, or optimized for the construction of gene microarrays on nylon membranes (Section 2), the probe preparation (Section 3), the hybridization of the arrays (Section 4), and the image analysis and data interpretation (Section 5). Higher density glass-based microarrays are likely to become readily available to more than a few specialized laboratories. Many of the considerations described here for nylon arrays will apply to glass arrays and are discussed (Section 6).

2 Gene array construction on nylon membranes

In this section, we provide recommendations—based on our experience—regarding choice of membranes, selection of test and control clones, DNA preparation, robotics, and spot density.

2.1 Nylon membranes

There are a number of commercially available nylon membranes, but careful testing reveals significant differences in terms of DNA binding capacity and uniformity. Hence, if one wishes to maximize sensitivity and reproducibility, it is critical to validate a membrane type in the particular system used. This is readily done by spotting identical DNA samples at several positions on the membrane. After hybridizing and quantitating (see below), values should be similar at all positions. All protocols described herein are routinely performed on Schleicher and Schuell Nytran Plus membranes (Cat. No. 414655).

2.2 Selection of test clones

The clones to be chosen for the arrays will typically correspond to genes of interest to the specific projects. There are a number of practical considerations regarding the specific region of the gene to be used.

2.2.1 Similarity to other genes

If a clone of interest is highly related to other genes, it is likely that there will be some level of cross-hybridization that could obscure the true hybridization signal for that gene. To address this problem, a region of the gene that is less or not similar to other family members should be selected whenever possible (e.g. 3′ untranslated region).

2.2.2 Repetitive elements

If a clone insert contains repetitive elements, such as Alu repeats, it will likely give an artificially high hybridization signal on the array. While this may be blocked to some extent (see below), it usually complicates the analysis of the signal generated and should be avoided if possible.

2.2.3 Insert size

As hybridization signals are based partially on molar ratios rather than mass, larger clone inserts will require more material to achieve signals similar to smaller ones. Larger inserts also increase the likelihood of repetitive elements and similarities to other family members. Clone insert size will also impact on the production of DNA to be arrayed. Indeed, if bacteria are arrayed directly on the membranes or used to produce plasmid DNA prior to the actual arraying, a larger clone will usually produce less DNA, due to slower growth rates of bacteria with large plasmids. If PCR is used to produce inserts to be arrayed, larger inserts are more difficult to amplify and generally produce less material. On the other hand, inserts that are too short may require significantly different hybridization conditions and should also be avoided. In general, we have obtained excellent results with clones in the range of 300–1000 bp.

2.2.4 Position within a gene

The position of a clone insert within a gene may also impact the hybridization signal it produces; depending on the labelling protocol used (see below). If a clone insert is located at the 5′ end of a larger gene, and oligo(dT) is used to prime reverse transcription during probe preparation, there may be only a small amount of probe corresponding to the region of interest, and subsequently little or no hybridization signal generated against that clone. When using random priming in the course of reverse transcription, insert position within a gene becomes less important.

2.3 Controls

There are a number of possible controls for gene array analysis. There are four types of controls that should be considered, although the specific choice of controls will obviously depend largely on the experiments being performed.

2.3.1 Negative controls

Two types of negative controls should be included and placed in numerous positions throughout the array. The first is a sample containing no nucleic acids, but otherwise treated identically to all other samples. This element will be measured to determine background signal intensity. The second type of negative control will consist of gene fragments from five to ten different genes that are not expected to hybridize to the probes used. For example, if the array consists primarily of human clones, then plant clones could be used as negative controls. Ideally, the sequence of such negative controls should be thoroughly compared to all available sequences from the species from which the arrayed clones are derived. While it is unlikely that all sequences in the probe will be known in the near future, this comparison may detect unwanted similarities.

2.3.2 Positive controls

An ideal positive control is a clone containing a gene known to be differentially expressed between two or more tissue samples. Once arrays are analysed, such positive control should exhibit the expected change in expression. However, a positive control of this type is not always available prior to analysis as details of expression patterns are not always available for the tissues or cell types under study. A second type of positive control are the housekeeping genes (e.g. GAPDH, cyclophilin, β-actin) which are expected to be expressed at relatively similar levels in all tissues or cell types analysed. These controls serve as positive indicators of successful labelling and hybridization reactions and also have the potential for use in data normalization (see Section 2.3.4).

2.3.3 Sensitivity controls

It is possible to determine the sensitivity of hybridization with controls designed for this purpose. To this end, one starts with a clone that will not hybridize to the experimental probe, as described under negative controls (Section 2.3.1). The clone should be in a vector that contains a synthetic promoter such as T7 or SP6. The clone is then used to direct transcription of a synthetic RNA with appropriate reagents (e.g. Promega Riboprobe Systems, Cat. No. P1420). After quantitation, known amounts of this synthetic RNA are added to an aliquot of the RNA that will be labelled as probe. A single control will determine if a defined sensitivity has been reached, although several different clones may be used to measure a range of sensitivities. It is critical that the sensitivity control clone deposited on the array does not contain sequences that are the reverse complement of sequences present in the probe. For instance, if oligo(dT) is used to prime the labelling reaction (see Section 3.3.4), there should be no poly(A) sequences in the sensitivity control clone. It is also important that the synthetic RNA contains sequences that will be primed in the labelling reaction. Hence, when using oligo(dT) for labelling, the synthetic RNA will need a poly(A) tract at its 3' end. This is obviously not a concern when using random oligonucleotides during probe preparation. Alternatively, primers specific for the synthetic gene may also be added to the probe synthesis reaction.

2.3.4 Normalization controls

In some experiments, it may be desirable to normalize the hybridization data to the expression of specific known clones. It is known that the expression of many of the so-called 'housekeeping' genes varies between tissues (e.g. actin, ubiquitin). However, if there are genes whose expression is known to be constant between the experimental tissues, the hybridization signals on the arrays can be normalized to these controls. This approach can also be compared to the method described in Section 5.2.3.

2.4 **DNA preparation**

The source of DNA used for arraying will impact on the construction and application of the arrays. Clone-containing bacteria may be spotted directly on membranes, grown, and processed for rapid production of arrays. Purified DNA (plasmid DNA or PCR product) can be spotted directly onto nylon membranes and we found this to be the method of choice for the construction of reproducible arrays with low background.

2.4.1 Bacteria

Arrays of clone-containing bacteria have the clear advantage of being relatively low cost and very rapid to produce. When a simple yes/no answer is desired, such as crude expression profiling for particular clones, these are often the arrays of choice. However, for detailed and sensitive examination of gene expression, purified DNA will produce much more robust arrays. It should be noted that the overall clone density will also be lower with bacteria-based arrays.

2.4.2 Plasmid DNA

Many commercial kits are available today for the rapid purification of plasmid DNA in 96-well format. Our laboratory found the Qiagen R.E.A.L. Prep 96 System (Cat. No. 26171) to be extremely robust with a single investigator able to process four plates in 90 minutes. After purification, each of the plasmid DNA samples should be run on agarose gels with appropriate mass markers to verify that sufficient material is available for constructing the arrays. Alternately, DNA can be quantitated in microtitre format with an appropriate plate reader and reagents such as the Perspective BioSystems CytoFluor Multi Plate Reader 4000 (Cat. No. MIFSOC2TC) with the CytoProbe Hoescht DNA Assay Kit (Cat. No. CPDNA9610) or the Molecular Probes, Inc. Pico Green™ dsDNA Quantitative Reagent Kit (Cat. No. P-7589).

2.4.3 PCR amplification

As hybridization intensity is affected in part by the number of copies of a gene fragment present in a given spot on the array, PCR products offer an advantage over plasmids. Indeed, for any given mass of DNA, a PCR amplified insert will contain more gene copies than a corresponding plasmid. As relatively large quantities of materials are necessary to prepare arrays, it is valuable to optimize PCR reactions to go to completion. The following parameters may be optimized to increase yields for different vector and primer combinations. Our optimized procedure is described in *Protocol 1*.

i. PCR buffer

The buffer used for the PCR amplification will have a large impact on the amount of PCR product generated. We found that the tricine buffer described in *Protocol 1* gives reproducibly high yields with numerous templates and primers. The magnesium concentration will also affect yields and should be titrated to optimize yield while retaining specificity.

ii. PCR primers

Primer sequence and length can have a great impact on yields. It is suggested that several primers be designed and then tested with 24–48 representative clones. Typically, we have been using the same set of 'universal' primers in our PCR reactions as most or all or our clones reside in the same 'universal' plasmid. This is a convenient feature as it allows us to make a single PCR master mixture (buffer, primers, nucleotides, and enzyme) and then multiplex our PCR reactions in 96- or 384-well reaction plates. In those situations where a single PCR reaction plate will contain clones using different vectors, we have found that a PCR master mixture possessing a cocktail of up to three separate 'universal' primer pairs not to interfere with yields or specificity.

iii. Polymerase

Too little polymerase will lead to reduced yield, while excess polymerase will not increase yields but will increase cost. The polymerase can be titrated up until no increase in product is observed. Typically, titration against 24–48 representative clones is sufficient to identify the appropriate polymerase concentration. While we have observed no significant differences when using *Taq* polymerase supplied from different vendors, we do recommend the retesting of the optimized PCR conditions if and when you change suppliers.

iv. PCR template

Purified plasmid DNA or clone-containing bacteria may be used as template. Bacteria may be used from fresh or thawed cultures, and robotic pin tools may be used to transfer the bacteria into the PCR mix. The amount of bacteria added should be titrated up until no product increase is seen, or else a decrease in yields will be observed. Testing against 96 clones is recommended.

v. PCR amplification program

The thermocycling program used should be optimized for the primers chosen for amplification. Several annealing temperatures should be tested to determine the highest yields. Longer extension times are useful if there are mix insert sizes of the template clones. Typically, 24–48 representative clones are sufficient for testing these parameters.

2.5 Robotics

There are a number of robotic workstations now available that are capable of depositing samples into arrays on nylon membranes. It is important that the robot is accurate and consistent, as repeat over-spotting is typically used in the construction of arrays. Pin tools used to transfer liquid from microtitre plates to nylon membranes vary greatly in design and this will affect spot diameter as well as the amount of liquid transferred during spotting. This will in turn affect density limits and repetitions necessary for transfer. Testing of the pin tool and spot diameter can be determined by repetitive spotting of three to five control genes followed by hybridization with their ^{32}P or ^{33}P labelled cognate sequence.

Protocol 1

Amplification of clone inserts by PCR

Equipment and reagents

- Thermocycler
- $10 \times$ PCR buffer: 300 mM tricine, 25 mM MgCl$_2$, 50 mM 2-mercaptoethanol
- dNTPs (any vendor)
- *Taq* polymerase (any vendor)

Method

1. For a 100 μl reaction, combine the following:
   - 10 μl of $10 \times$ PCR buffer
   - 10 μl of 2 mM each dATP, dCTP, dGTP, dTTP
   - 50 pmol of each primer, forward and reverse
   - 2.5 U of *Taq* polymerase
   - Water to 100 μl
2. Inserts are typically amplified with the following thermocycling program:
   (a) One cycle of 94°C for 4 min.
   (b) 38 cycles of 94°C for 30 sec, 50°C for 30 sec, 72°C for 3 min.
   (c) One cycle of 72°C for 10 min.

Most robots work with either 96- or 384-well microtitre plates or both. The choice of plates will depend in part on the number of samples. 384-well plates offer a number of advantages. First, since fewer robotic manipulations are required, array construction will be faster and potentially more uniform. Secondly, the wells of 384-well plates are smaller when compared to 96-well plates. This allows the preparation of less material for spotting since the smaller volumes will still be sufficient for proper deposition. Obviously, whichever plate is chosen, it is essential that it be compatible with the spotting robot.

A variety of robotic workstations and pin configurations have been used in our laboratory to generate high quality arrays. These workstations have included the BioMek 2000 (Beckman), the 'Q'Bot (Genetix Ltd.), and the Flexys™ (Genomic Solutions Inc.). We describe our optimized procedures for use with the Flexys™ robotic workstation in *Protocol 2* for DNA and *Protocol 3* for bacteria.

2.6 Density

The density of clones in the array will be determined by several factors.

2.6.1 Clone number

If there is a large number of clones to be spotted, either a higher density or larger surface will be required. In some instances it may be necessary to go to multiple membranes.

2.6.2 Spot size

A typical spot size on nylon membrane is approximately 900 μm in diameter. This size limits spot density to 1000 μm spot-to-spot centres or 100 clones/cm^2. Hence, on an 80 $\times$ 120 mm membrane, the maximum number of arrayed clones is 9600 or twenty five 384-well plates.

As discussed in Section 2.4.1, fewer clones should be used if the DNA source is clone-containing bacteria. In this case, the practical upward limit is closer to nine 384-well plates or 3456 clones. The actual density of colonies will vary depending on strains used, insert sizes, and growth conditions.

2.6.3 Detection system

Spot density can be affected by the method used to determine hybridization signal, and by the strength of the signals generated (see Section 5). For higher density work, it is important to determine empirically the maximum usable spot density.

Protocol 2

Deposition of DNA onto nylon membranes

Equipment and reagents

- Flexys™ robotic station (Genomic Solutions Inc.)
- Stratalinker (Stratagene)
- Blotting paper (Schleicher and Schuell GB002)
- Whatman 3MM paper (Whatman, International Ltd., Cat. No. 3030-917)

- Nylon membrane (Schleicher and Schuell Nytran Plus, Cat. No. 414655)
- Denaturing solution: 0.5 M NaOH, 1.5 M NaCl
- Neutralizing solution: 0.5 M Tris pH 7.5, 1.5 M NaCl

Method

The volumes stated are for use with the Flexys™ robotic station and are sufficient for 12 complete spotting runs of eight filters each. Volumes may be reduced for fewer filters, and may change with different systems. All volumes should be determined empirically for a specific system.

1. For each clone, DNA should be at a starting concentration of approx. 300 ng/μl of either plasmid DNA or amplified insert. Transfer the DNA samples to a new 384-well microtitre plate. To 7 μl of DNA (300 ng/μl), add 14 μl denaturing solution. Note that the DNA should be denatured just prior to array construction and discarded after use.

2. Place the nylon membranes on two pieces of blotting paper and then in the robot according to the manufacturer's instructions. With the Flexys™, we place the membrane (80 $\times$ 120 mm) and paper in the lid of a microtitre plate. To do this, the blotting paper is first placed into the microtitre plate lid, saturated with denaturing solution, and excess solution drained off. The nylon membrane is then placed

gently on top of the paper with care taken not to introduce any air pockets. DNA is deposited onto the membrane with the pin tool according to the manufacturer's instructions. The spotting is repeated four more times to increase the amount of material deposited. It is necessary to increment the number of over-spottings by one every two filter sets, to account for the decreased volume of DNA solution.

3. After spotting, place the nylon membranes on Whatman 3MM paper soaked in neutralizing solution for 5 min.

4. Repeat the neutralizing step once.

5. Briefly rinse the membranes in distilled water to remove salts.

6. Air dry the membranes on a new piece of Whatman 3MM paper.

7. UV cross-link DNA to the membranes using the auto cross-link setting on a Stratagene Stratalinker or equivalent (120 000 μJoules/cm^2).

8. Store the cross-linked membranes in a sealed bag, frozen at $-80\,°$C or desiccated.

Protocol 3

Deposition and processing of bacteria on nylon membranes

Equipment and reagents

- See *Protocol 2*
- Carbenicillin (Sigma, C1389)

Method

1. Clone-containing bacteria to be spotted should be fresh overnight cultures in standard growth media (typically LB plus carbenicillin, 100 μg/ml final concentration).

2. Nylon membranes should be placed on approximately seven pieces of blotting paper. More blotting paper will result in larger colony size. The exact number of blotting paper will depend on the bacterial strain used and the desired spot density, and should be empirically determined.

3. For the Flexys™, membranes and blotting paper are placed into microtitre plate lids and saturated with growth media as described in *Protocol 2*, step 2.

4. Bacteria should be spotted onto the membranes once. The pin tool must be sterilized between each plate. This protocol will vary between robots. For the Flexys™, use 5 sec in water followed by 5 sec in 70% ethanol, 5 sec in 95% ethanol, and then a drying step as per manufacturer's instructions.

5. Following spotting, cover with a second microtitre plate lid and grow overnight, face up, at 37$\,°$C until colonies are clearly visible, but not large (approx. 1 mm diameter).

6. Transfer the membranes to Whatman 3MM paper that is saturated with denaturing solution for 5 min.

Protocol 3 continued

7. Move the membranes to a dish filled with neutralizing solution. Once submerged, the colonies should be gently scraped off with a gloved hand.

8. Transfer the membranes to Whatman 3MM paper saturated with neutralizing solution for 5 min.

9. Briefly rinse the membranes with distilled water to remove salts.

10. Air dry membranes on a new piece of Whatman 3MM paper.

11. UV cross-link DNA to filters using the auto cross-link setting on Stratalinker or equivalent (120 000 μJoules/cm^2).

12. Store membranes in a sealed bag, frozen at $-80\,°C$ or desiccated.

3 Probe preparation

There are a number of points to consider in the preparation of probes for hybridization of microarrays from RNA selection and purification to the actual probe synthesis. Simple probes, such as a labelled insert or a single *in vitro* transcribed RNA, are relatively easy to work with in screening applications. Complex probes, such as those derived from RNA isolated from a tissue, are more challenging in their use. *Protocols 4* and *5* have been optimized for use with complex probes, but will also work well with simple probes.

3.1 RNA selection

The scope of the experiments will determine which RNAs are of interest for probe synthesis. It is highly critical that the tissues or cells used for RNA isolation be of the best quality possible. One can use RNAs from species other than that of the genes contained on the array, e.g. RNAs from rodent tissues—easier to obtain at very high quality—can be used to make probes for hybridization against human gene microarrays. In these cases, it may be necessary to change hybridization conditions to lower stringency to obtain useful results. The interpretation of these 'cross species' experiments is somewhat different and is further discussed in Section 5.2.6.

3.2 RNA purification

The quality of the template RNA (e.g. integrity and purity) will dramatically affect the hybridization results obtained by microarray analysis. We recommend that that the integrity of all RNA samples be inspected electrophoretically with those samples exhibiting degradation re-isolated. Our experience also shows that RNA must be of extremely high purity to ensure both the reproducibility and sensitivity of the final data. *Figure 1A* illustrates the effect of low purity in microarray analysis. There are several commercially available RNA isolation kits that will yield different results depending on the quality and amount of starting materials. In general, it is best to use the minimum amount

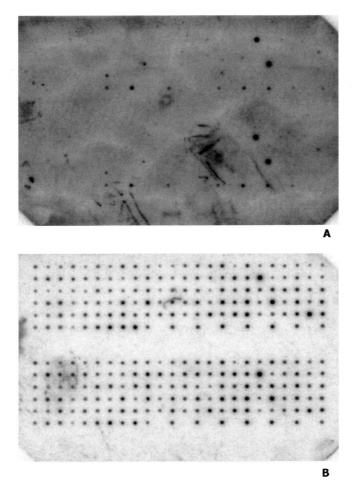

Figure 1 The effects of RNA purity on microarray analysis on nylon membranes. (A) Arrays on nylon were made as described in *Protocol 2*. RNA was obtained from Clontech (cerebellum, Cat. No. 6543-1) and not subjected to any further purification. RNA was labelled as described in *Protocol 5* using dCT^{33}P. Hybridization was in a sealed box for 16 h as described in *Protocol 6*. After washing, the membrane was exposed to a PhosphorImaging screen and then scanned with a Storm PhosphorImager (Molecular Dynamics). (B) The same procedures were used as in (A) with a duplicate nylon array, except that the RNA was first passed through a Chromaspin 30 column (Clontech, Cat. No. K1321-1) following the manufacturer's protocol.

of tissue or cells recommended by the manufacturer. In our hands, the Qiagen RNA–DNA Kit (14142) provides RNA of exceptional quality. It is also possible to further purify RNA of low purity to improve its performance in microarray analysis as shown in *Figure 1B*.

It is usually not necessary to isolate poly(A)$^+$ RNA for use with microarrays. In some cases, such as bacterial RNAs, this will not even be possible. If already available, poly(A)$^+$ RNA makes an excellent template for probe production.

When using total RNA, we recommend that labelling reactions be primed with oligo(dT) whenever possible (see Section 3.3.4).

Protocol 4

Isolation of total RNA using the Qiagen® Tip 100 RNA–DNA Kit

Reagents

- Qiagen® Tip 100 RNA–DNA Kit (Cat. No. 14142)
- Qiashredders™ (Qiagen, Cat. No. 79654)

Method

Begin with NO MORE THAN 100 mg tissue or 2×10^7 cells per column. Multiple columns may be used to isolate RNA from additional tissue. Add 10 μl 2-mercaptoethanol to each 1 ml of buffer QRL1 used. Pre-heat buffer QRU (with urea added) to 45 °C. Heat a water-bath to 60 °C. Buffers QRL1 and QRU (and QRV1, QRV2, QRE, QRW, QRU) are provided by the manufacturer.

1. (a) Tissue. Add 2 ml buffer QRL1 to tissue and then homogenize tissue with rotor-stator homogenizer for 30 sec.

   (b) Cells. Pellet cells by centrifugation at 500 g for 5 min and discard supernatant. Add 2 ml buffer QRL1 and mix thoroughly by pipetting up and down until pellet is completely lysed. Homogenize by conventional homogenizer for 15 sec or by splitting into three Qiashredders™ and centrifuging for 2 min at maximum. Pass lysates through the same Qiashredders™ two more times and then pool.

2. Add 2 ml buffer QRV1, mix well by inversion, and centrifuge 20 min at 15 000 g at 4 °C.

3. Carefully transfer supernatant to new tube, add 3.2 ml isopropanol, and mix well by inversion.

4. Incubate on ice for 5 min, centrifuge 30 min at 15 000 g at 4 °C, and discard supernatant.

5. Resuspend pellet in 1 ml QRL1 by heating for 3 min at 60 °C and then vortexing. Repeat heating and vortexing twice.

6. Mix sample with 9 ml QRV2 and centrifuge 5 min at 5000 g at 4 °C.

7. In the meantime, equilibrate the Qiagen column by adding 3 ml buffer QRE and allowing to drip.

8. When column stops dripping, add supernatant from step 6.

9. When column stops dripping, wash with 12 ml buffer QRW.

10. When column stops dripping, change collection tubes and elute RNA by adding 6 ml pre-heated buffer QRU.

11. Add 6 ml ice-cold isopropanol to sample and mix well by inversion.

12. Incubate on ice for 10 min then centrifuge 30 min at 15 000 g at 4 °C.

13. Carefully remove supernatant and wash pellet with 2 ml of 70% ethanol.

14. Centrifuge 15 min at 15 000 g at 4 °C.

15. Discard supernatant and allow pellet to air dry for 10 min.

16. Resuspend pellet in 125 μl RNase-free water. Quantitate RNA by measuring absorption at 260 nm. Typical yields are 125 μg total RNA/column/100 mg of tissue.

3.3 Probe synthesis

The probe is generated by reverse transcribing RNA in the presence of priming oligonucleotides and a labelled nucleotide analogue. There a several aspects of producing labelled probe that will affect the overall analysis. An optimized probe preparation protocol is given in *Protocol 5*.

3.3.1 Reverse transcriptases

There are a number of commercially available reverse transcriptases. In our hands, comparisons show that probes made with LTI's Superscript II (Life Technologies, Cat. No. 18064-014) routinely produce stronger hybridization signals (data not shown).

3.3.2 Nucleotide concentration

The ratio of labelled to unlabelled nucleotide will affect the specific activity of the probe. With radiolabelled nucleotides, it is suggested that an equal amount of labelled and unlabelled nucleotide be added as the concentration of the labelled nucleotide is relatively low and will limit the amount of product synthesized. If other labelling systems are used (e.g. fluorescent dyes), the specific activity of the probe should be determined by hybridization. This is necessary as large moieties attached to nucleotides and incorporated into probes can affect hybridization efficiency. By titrating labelled and unlabelled nucleotides, it is possible to optimize for the strongest hybridization signals.

3.3.3 Radiolabel

The type of radioisotope used to detect and quantitate hybridization signals will affect the sensitivity of the analysis, the dynamic range (maximum spot intensity to background), and the upper limit of spot density. ^{32}P-labelled nucleotides provide excellent sensitivity and a greater dynamic range due to their relatively high energy. However, this higher energy produces a larger signal radius for a given number of atoms, when compared to ^{33}P (please see *Figure 2* in plate section). To avoid this influence of ^{32}P signal from neighbouring clones, it is necessary to spot at a relatively low density. Clones may be spaced much closer when ^{33}P is incorporated into the probe. While ^{33}P will not achieve the ultimate sensitivity and dynamic range possible with ^{32}P, it will allow a more rapid analysis for larger number of clones.

Other, non-radioisotopic, methodologies are available to detect and quantitate hybridization signals on nylon arrays. Fluorescent probes are inadequate due to the intrinsic fluorescence of the nylon membrane. However, the switch from nylon to silica or glass microarrays allows the use of fluorescent probes and is discussed in Section 6. Chemiluminescent methodologies such as the Photo-gene™ Nucleic Acid Detection System (Life Technologies, Cat. No. 30062), the BrightStar™ Psoralin Biotin Kit (Ambion, Cat. No. 1480), and the Illuminator™ Chemiluminescent Detection System (Stratagene, Cat. No. 18192-057), while not tested extensively in our laboratory for use on arrays due to our success with radiolabels, are available and may provide satisfactory and reproducible results. *Protocol 5* has been optimized for use with ^{33}P.

3.3.4 Primers

The reverse transcription reaction is primed using synthetic oligonucleotides. Oligo(dT) may be used to prime only those messages with poly(A)$^+$ tails. Random primers will prime on all sequences, and should be used when there are no poly(A)$^+$ tails, e.g. for microbial RNA samples. While hexamer oligo-nucleotides are commonly used for such random priming, we have obtained more consistent results with random nonamers (data not shown). When prob-ing a limited number of clones, it is possible to use gene-specific primers. As each primer type (e.g. oligo(dT), hexamer, nonamer, or gene specific) may result in a different signal intensity and hybridization pattern, depending on which clones are chosen (see Section 2.2), it is necessary to use the same primers for all experiments that will be compared.

3.3.5 Probe purification

After the labelling procedure, it is necessary to remove unincorporated label and primers. There are many commercially available products for this purpose. Again, these can be tested to determine which is most effective in a given microarray system. We found that the Chromaspin 30 columns (Clontech) result in very reproducible removal of unwanted materials.

Protocol 5

Probe production by reverse transcription with ^{33}P

Reagents

- LTI Superscript II PreAmplification Kit (Life Technologies, Cat. No. 18089-011)
- dNTP mix: 10 mM each dATP, dGTP, dCTP, and 16.5 μM dCTP
- dCT^{33}P (Amersham Pharmacia Biotech, Cat. No. AH.9905)
- Chromaspin 30 column (Clontech, Cat. No. K1321-1)

Method

This protocol is for use with the LTI Superscript II PreAmplification Kit.

1. Add RNA (20 μg total or 0.5 μg poly(A)$^+$), 2 μg of oligo(dT)$_{12-18}$, and DEPC treated water to a final volume of 7 μl. (Random oligonucleotides may also be used. Add 375 ng of random nonamers.)

2. Heat at 70°C for 10 min, and then place on ice for 1 min.

3. Add 2 μl of 10 × PCR buffer, 2 μl of 25 mM MgCl$_2$, 1 μl dNTP mix, and 5 μl dCT^{33}P. Mix by pipetting up and down.

4. Incubate mixture for 5 min at 42°C.

5. Add 1 μl SuperScript II enzyme (200 U/μl). Mix by pipetting up and down.

6. Incubate for 50 min at 42°C. Heat at 70°C for 15 min.

7. Add 1 μl of *E. coli* RNase H (2 U/μl) and incubate for 20 min at 37°C.

8. Add 29 μl of water and spin through a Chromaspin 30 column using the recommended protocol.

9. Count 2 μl in a scintillation counter to determine activity. Specific activity of the probe should be at least 2.5×10^6 c.p.m./μg total RNA or 1×10^8 c.p.m./μg poly(A)$^+$ RNA.

4 Hybridization

In this section we provide recommendations—based on our observations—for hybridization conditions, buffers, and signal detection. It is important to note that the various factors described below have been optimized for use with *Protocols 5* and *6*.

4.1 Hybridization conditions

After depositing a gene array onto a nylon membrane, it is hybridized with labelled probe. The hybridization conditions and subsequent handling will greatly impact the quality of the final data. *Figure 3* (please see plate section) shows the results of successful nylon microarray hybridization and the hybridization procedure is described in *Protocol 6*.

4.1.1 Hybridization buffer

It is important to use a buffer system that produces uniform backgrounds of low intensity. Pre-hybridization can reduce background levels even further. Hybridization buffers should be tested with specific membranes for background uniformity and level. Our recommended buffer system is given in *Protocol 6*.

4.1.2 Repetitive sequences

As discussed earlier, clones that contain repetitive sequences may show high intensity signals. Unlabelled low complexity DNA (Cot-1 DNA) may be added to the hybridization buffer to reduce these signals. The amount of this low complexity DNA used should be titrated as large quantities of unlabelled DNA can reduce overall hybridization signals.

4.1.3 Probe concentration

Excessive amounts of probe will increase background without increasing desired signals. Alternatively, too few counts will decrease the desired signal, making real signal difficult to discriminate from background. We recommend starting with 1×10^6 c.p.m./ml of hybridization buffer. This may be titrated with the specific membrane/buffer systems used.

4.1.4 Time and temperature

Hybridizations with complex probes should generally proceed for approximately 16 hours to ensure that messages present in low quantities will hybridize (18). It is important that experiments that are to be directly compared—e.g. comparison of normal versus diseased tissues—be hybridized for the same time and temperature to ensure that the hybridization results are comparable. The hybridization and wash temperatures will depend on the specific buffers used.

4.1.5 Detection system

The detection systems in use for nylon microarrays hybridized with radio-labelled probes are autoradiography films and PhosphorImagers. Autoradiography films are relatively standard in most laboratories and are useful when small numbers of clones are under study and only simple 'on/off' questions are being addressed. However, when hundreds or thousands of clones are under study and quantitative results are required, the use of a PhosphorImager greatly simplifies the process. Hybridized membranes should be exposed to PhosphorImaging screens for 1–72 hours, depending on the specific activity of the probe and total counts detectable on the membrane itself. Following exposure to the hybridized membrane, the PhosphorImaging screen is processed according to the manufacturer's recommendation.

Protocol 6

Hybridization of nylon microarrays

Reagents

- Cot-1 low complexity DNA (Life Technologies, 15279-011)
- $20 \times$ SSC: 3 M NaCl, 0.3 M sodium citrate pH 7.0

- Hybridization buffer (Church buffer) (19): 0.5 M Na_2HPO_4 pH 7.2, 7% SDS, 1% BSA (fraction V), 1 mM EDTA

Method

All membranes should be handled as little as possible and only by the edges using forceps.

1. Place membranes in appropriate container (air-tight boxes reduce the amount of membrane handling. Rotating bottles or heat-sealed bags can be substituted).

2. For an 8×12 cm membrane in a 9×15 cm box, use 12 ml of hybridization buffer for pre-hybridization. 100 ng/ml of Cot-1 low complexity DNA may be added if there are clones containing repetitive elements. For two membranes, place back to back and use a total of 15 ml hybridization solution to ensure that they are completely covered. Greater than two filters per box is not recommended. Gently agitate the filters for 1 h to overnight at 65 °C (1 h pre-hybridization is sufficient).

Protocol 6 continued

3. Prepare the same volume of hybridization buffer used for pre-hybridization. Use 1 $\times$ 10^6 c.p.m. of radiolabelled probe for each 1 ml of hybridization buffer. Boil the purified probe for 3 min and then place on ice for 1 min. Add probe to 1 ml of fresh hybridization buffer, mix, and then add to remaining buffer. (Avoid boiling the probe with an aliquot of the Church buffer as incubation on ice results in severe SDS precipitation.)

4. Remove the hybridization buffer from the pre-treated membranes and add fresh hybridization buffer with the denatured radiolabelled probe.

5. Gently agitate at 65°C for approx. 16 h to ensure that messages present at low levels are hybridized.

6. Wash membranes in 2 $\times$ SSC at 65°C for 15 min with gentle agitation.

7. Wash membranes in 0.2 $\times$ SSC, 0.1% SDS at 65°C for 15 min.

8. Seal membranes on Saran Wrap or equivalent plastic film, and expose to recently erased PhosphorImaging screens as per manufacturer's instructions.

9. Expose screen for 1–72 h depending on the specific activity of the probe and total counts detectable on the membrane.

10. Scan PhosphorImaging screens as per manufacturer's instructions and process the data as discussed in Section 5.

5 Image analysis and data interpretation

This section describes methods and considerations for the analysis and interpretation of array data. At the time of this writing, there exists no accepted standard for data quantitation, normalization, or quality scoring. While this lack of standardization limits one's abilities to directly compare data derived from other laboratories, it is not inhibitory to the generation and use of array data and internal databases.

5.1 Data quantitation

Once the microarray image has been captured electronically, it is necessary to quantitate the intensity of hybridization at each position on the array. While most PhosphorImagers come with generalized software designed to capture fluorescent signals, a program specifically designed for this purpose best accomplishes quantitation of arrays. A commercially available program designed for analysis of microarrays on nylon membranes is *ArrayVision*™ by Imaging Research (http://imaging.brocku.ca). This program provides a sophisticated grid placement method, tools for background correction, and image filtration. An example of grid placement is shown in *Figure 4* (please see plate section).

5.2 Data interpretation

The analysis and interpretation of array data can be very complex. To minimize variation and reduce complexity, it is important that hybridization experiments be done with the same protocols, conditions, and reagents whenever possible. We discuss here various points to consider to facilitate data interpretation.

5.2.1 Spot intensity

The hybridization signal intensity for a given spot on the array is affected by a number of factors in addition to the amount of transcript actually represented in the labelled probe. Factors such as location of the clone insert within the gene coupled with the primers used in probe synthesis, GC content, presence of repetitive elements, and amount of DNA deposited on the membrane will all contribute to the final signal detected for each clone. These variable factors make comparisons of expression between clones on the same membrane difficult. Arrays are best used to compare signals for a given clone on replica membranes hybridized with different probe samples to look for changes in expression between two or more tissues, cells, or conditions.

5.2.2 Duplicate experiments

By conducting array experiments in duplicate, one can increase the confidence in the data generated. Arrays can be duplicated on one filter, or on a second, separate filter. When using two separate filters, they can be hybridized simultaneously in the same chamber; data can be collected from each array and then compared. A scatter plot, such as the one shown in *Figure 5* (please see plate section), is useful to rapidly visualize the quality of duplicate data. The elemental display function in the *ArrayVision*™ software can also be used to determine variation between two filters. For example, our system has been optimized and validated such that duplicates vary by less than 1.5-fold. Such reproducibility not only increases the confidence in the data generated, but also provides the baseline for which differences in expression between to two or more tissues, cells, or conditions are deemed significant. Duplicates that vary widely should be examined more closely. Successful duplicates can be averaged for use in further analyses.

5.2.3 Normalization

To compare experimental data between different probes, it is usually necessary to normalize the raw counts obtained after initial analysis of the arrays. Normalization will allow comparisons between experiments where the specific activity of the probes or exposure time is different. A simple approach to normalizing data is incorporated into the *ArrayVision*™ software. This procedure divides the raw value for each spot by the sum of values for all spots and then multiplies by the number of elements in the array. The intensity for each spot is then reported as its proportion of total expression. This method works very well for comparing samples that are roughly similar over the genes arrayed and when a good percentage of the arrayed clones hybridize with the labelled probe. If the distribution varies greatly between samples, then this process will introduce

Normalization

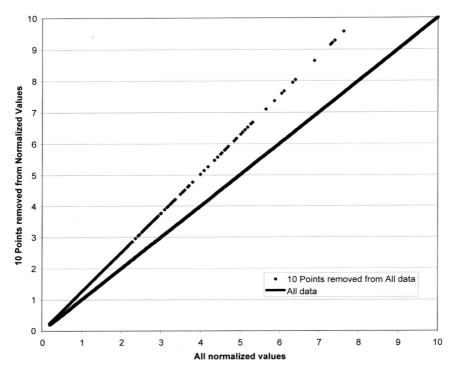

Figure 6 Normalization of expression values. Data from *Figure 3A* were subjected to the normalization program from the *ArrayVision* software. To simulate a different expression profile, the ten highest raw values from quantitating the array in *Figure 3A* were set to zero. All raw values were then normalized as described in Section 5.2.3 and compared to the original normalized values from the array. While only 0.65% (10/1536) of the elements were changed, individual normalized values were changed by approximately 25%.

some error. *Figure 6* shows the effects of removing ten spots prior to normalization. If the distributions are severely different, it may be necessary to use other normalization methods (for examples, see refs 20 and 21).

5.2.4 Background considerations

Background signals can affect the intensity of any given element in an array. The uniformity of background will in part determine the sensitivity of detection possible on an array. It may be possible to increase this sensitivity by accounting for local background surrounding an element. The *ArrayVision*™ software provides several different methods for background determination and subtraction. The methods can be tested by using each one on the duplicate filters. The most appropriate method will be the one with which most duplicates agree very closely.

5.2.5 Identifying expression patterns of interest

When comparing several hybridized arrays, it is possible to look for complex expression patterns such as co-regulation by induction or repression of transcription. When comparing two experiments, one is typically interested in finding expression levels that change either up or down. Once all of the data have been collected, a scatter plot will provide a rapid visualization of the overall differences between the tissues or conditions (please see *Figure 7* in plate section). On the other hand, the *ArrayVision* software provides an elemental display to show differences between two arrays. This display is based on the data generated and the settings used in the program. An example of such display is shown in *Figure 8* (please see plate section). Values can also be generated by calculating the ratio of the average values from one experiment to another. The spots that show the most change can be quickly identified by sorting these values. Another option for such analyses is the software package, *Spotfire Pro II* (IVEE Development AB). This is visualization tool that allows for interactive exploration and mining of databases and datasets.

5.2.6 Cross species analysis

When an RNA probe is used from a species other than the one of the genes used to produce the array, the interpretation of the data is somewhat different. The initial steps are done as described above and genes will be identified that show differences in expression between probes. These changes may correspond to the true homologs from the species being tested (often referred to as 'orthologs'), or to other genes that share sequence homology with the clones used on the array (for example, related gene family members often referred to as 'paralogs'). Comparison of gene sequences across species may give an indication as to which possibility is relevant for a given spot, but this is not always possible due to the currently limited amount of sequences in public databases for species other than human. In some cases, it will be necessary to follow up with other experiments to determine exactly which gene is changing in the test species.

3 Construction and use of DNA microarrays on glass

Arrays of DNA samples on membranes as described above have practical limitations when attempting to analyse the expression patterns of several thousands of genes. For example, the porous nature of the membranes themselves drastically limits the number of clones that can be placed on a single surface. In addition, membranes are typically 80 × 120 mm in size, require millilitre volumes of hybridization buffer, and are limited to radioactive, chemiluminescent, and colorimetric hybridization detection methods due to autofluorescence and scattering (22). While these detection methods are sensitive, most radioactive and chemiluminescent signals do disperse and hence affect not only the total density of the array but also its resulting analysis.

In contrast, microarrays constructed on glass surfaces are compatible with fluorescently labelled probes. As fluorescent probes exhibit little to no signal dispersal, they allow for very dense array spacing. Appropriately treated glass permits the deposition of DNA samples to at least 100 μm spot diameters, thus, 10 000 discrete DNAs can be placed in a 1.0 cm^2 array (23). In addition, appropriately treated glass surfaces exhibit very low non-specific binding of hybridization probes which results in considerably lower backgrounds than seen with nylon membranes. A result of restricting the spotting area to a small region on a non-porous surface is that the volumes required for hybridization are much smaller. Thus, for hybridizations involving complex probes, the minimal volumes required for hybridization of the microarrays make it practical to achieve high probe concentrations.

The protocols described above for nylon arrays are compatible with glass microarrays following slight modification. The considerations below are relevant to glass microarray systems where pre-synthesized DNA samples are applied to the glass surface through physical spotting (8) rather than *in situ* synthesis methods involving photolithography or physical masking (7).

6.1 Glass selection

The most common substrate used are surface-derivatized, non-fluorescent, white glass slides. Commercially derivatized glass slides are available (e.g. poly-L-lysine coated, silylated, silanated) and have been used successfully. In addition, the interested investigators can derivatize their own glass slides with a minimum of effort (`http://cmgm.stanford.edu/pbrown/array.html`). It should be noted that the choice of derivatization and its associated DNA attachment chemistry greatly impact upon DNA sample preparation and both pre- and post-spotting manipulations.

Affymetrix and Synteni/Incyte currently market pre-fabricated microarrays with additional companies expected to enter the fray (e.g. Clontech, Genome Systems, and Research Genetics). As the options for purchasing pre-fabricated arrays continue to increase, the 'technology' will become more affordable.

6.2 DNA source

Purified PCR amplicons ranging in size from 300–1000 bp provide the best source of DNA samples to be spotted (see *Protocol 1*). The purification methodology depends on the type of glass derivatization and the DNA attachment chemistry. Purification methodologies can vary from simple ethanol precipitation to column chromatography. The final concentration of the DNA samples is also dependent upon the chosen glass chemistries, hence final spotting concentrations must be determined by the investigator. In all such tests, it is strongly advised that actual hybridization conditions be employed.

6.3 Microarray robotics

At the time of this writing, few robotics systems are commercially available for the construction of glass microarrays. Known vendors of arrayers are Molecular

Dynamics, Genetic Microsystems, Intelligent Automation Systems, Gene Machines, and Cartesian. In addition, both Synteni/Incyte and Affymetrix will manufacture custom chips for a fee. Details on current products and services have recently been reviewed (24–26). For the do-it-yourselfer, Pat Brown provides a detailed part list and instructions for the manufacture of his robotic system (`http://cmgm.stanford.edu/pbrown/array.html`).

6.4 Microarray density

Clone density is dependent upon the chosen robotic system and the glass/DNA attachment chemistries. Current systems are able to manufacture arrays at a density of 10 000 spots per cm^2 (23).

6.5 Probe preparation

As with nylon arrays, probe preparation is critical for hybridization of glass microarrays. Typically, probes are generated by reverse transcribing RNA in the presence of priming oligonucleotides and a fluorescently labelled nucleotide analogue. The most common analogues in use are those labelled with the Cy3 or Cy5 cyanine dyes manufactured by Amersham (Cat. No. PA53021 and PA55021). Other analogues, such as the Alexa dyes by Molecular Probes can also be used but care must be taken to match the appropriate excitation and emission spectra to the array scanner. In addition, methods for probe purification will need to be tested for each of the analogues chosen.

6.6 Hybridization and detection

In this section, we provide considerations for glass array hybridization conditions and detection systems.

6.6.1 Hybridization conditions

Hybridization and wash conditions are relatively simple and have been described in detail elsewhere (8, 23). In general, hybridization times range from 4–18 hours with washing lasting as little as 10 minutes. The necessity for reducing hybridization backgrounds by blocking or pre-hybridization steps is dependent upon the glass derivatization chemistry chosen and will need to be determined empirically.

6.6.2 Detection systems

There are a number of commercially available fluorescence detection systems for glass-based microarrays. CCD-based systems are available from Genomic Solutions™ and Vysis. The more widely used confocal laser scanners are available from Molecular Dynamics, General Scanning, and Genetic Microsystems to name a few. As with the arrayer, the do-it-yourselfer can also consult Pat Brown's web page.

6.7 **Data capture**

The commercially available detection systems are provided with software for quantitating the intensity for each spot of the array. In addition, companies such as BioDiscovery and Imaging Research have developed and are marketing independent software packages for the analysis of microarrays.

7 **Conclusion**

EST and genomic sequencing projects have already identified the coding sequence of every gene in several model organisms (27–30). Recent advances in nucleotide sequencing technologies promise the identification of the entire repertoire of human gene sequences by the year 2003. This knowledge cries for the development of technologies that can analyse in parallel the expression patterns of all genes from one organism at once. Current microarray technologies can readily address the needs for organisms with smaller genomes (14) and have the potential to be applied to the analysis of all human coding regions as well.

Although the potential is there, a number of limitations remain to be addressed. For example, as the sequence databases and the resulting clone collections become larger, there will be a need for the development of confirmed, validated, and reliable clone sets for each of the organisms under study (e.g. a continued expansion of the human Unigene set). At present, the researcher is faced with either sequence confirming the publicly obtainable clones or relying on the annotations provided by the various suppliers. The importance of knowing with certainty the identity of all clones on the arrays and the need to construct fully sequence-validated clone sets available for the scientific community at large cannot be understated. Array interpretation is only as good as both the numbers of genes represented and the quality and reliability of the sequence validation of the gene set used.

Many investigators struggle with the need for accurate measurement of small changes in gene expression. We have validated our array systems for the detection of less than 1.5-fold changes in gene expression from single experiments. By over-sampling or repeated experimentation, smaller changes could be obtained. However, this is difficult when signal intensities are weak due to the low abundance of the gene product of interest. Technological and methodological advances are and will continue to address this problem. Already more sensitive scanners and better fluorescent dyes are being developed to improve the detection sensitivities. In addition, many groups, including us, are developing better and more sophisticated labelling methodologies designed to detect low abundance genes. Alternatively, one can use other methodologies such as TaqMan™ (Perkin Elmer Cetus) to monitor small changes in expression of a handful of low abundance genes.

Microarray-based assays easily provide several hundreds of data points for thousands to even ten thousands of genes. Stand-alone, fully automated, and standardized software systems for the collection, quality scoring, and tracking of these data points is needed. Additionally, within this immense amount of data exists a massive amount of biological information that must be extracted. Many pharmaceutical and biotechnology companies are developing their own tools to organize and explore the meaning of the information contained within the array data. Mike Eisen *et al.* (27), have recently described a system of cluster analysis to arrange and visualize genes according to their similarities in gene expression patterns. This type of analyses is definitely a step in the right direction, but it is difficult to expect visualization tools to handle the millions of expression measurements that arrays can generate. As more sophisticated tools for data interrogation are developed, the array technologies will begin to realize their full potential.

In addition to providing broad and in-depth information on gene expression patterns in normal and diseased tissues—both in human and animal systems—gene microarrays are a powerful tool to help dissect the mechanism of action of drugs and drug candidates. Microarrays will also increasingly contribute to the analysis of metabolic pathways for drugs, the understanding and prediction of toxic or adverse events in response to drugs, as well as the potential identification of surrogate markers to follow the effect and dose of a drug in the clinical setting. The number and variety of microarray applications is virtually unlimited. For that reason, it will be absolutely critical that strict and broadly accepted quality standards be developed in the near future, so that data from different laboratories can be compared and even combined. The rush for applying this exciting technology should not leave quality assurance in the dust because low quality data will only generate poor biological conclusions. Finally, microarrays are not a panacea for gene expression analysis. Indeed, high-throughput methods for the analysis of differential expression at the protein level, a discipline called 'proteomics', are being developed in earnest and will be highly complementary to microarrays (31). In addition, the labour-intensive *in situ* hybridization and immunocytochemistry methodologies will continue to make important contributions through their vital link to histology and cytology.

References

1. Liang, P. and Pardee, A. B. (1992). *Science*, **257**, 967.
2. Welsh, J., Chada, K., Dalal, S. S., Chang, R., Ralph, D., and McClelland, M. M. (1992). *Nucleic Acids Res.*, **20**, 4965.
3. Hubank, M. and Shatz, D. G. (1994). *Nucleic Acids Res.*, **22**, 5640.
4. Diatchenko, L., Lau, Y. C., Campbell, A. P., Chenchik, A., Moqadam, F., Huang, B., *et al.* (1996). *Proc. Natl. Acad. Sci. USA* **93**, 6025.
5. Weinstock, K. G., Kirkness, E. F., Lee, N. H., Earle-Hughes, J. A., and Venter, J. C. (1994). *Curr. Opin. Biotech.*, **6**, 599.
6. Velculescu, V. E., Zhang, L., Vogelstein, B., amd Kinzler, K. W. (1995). *Science*, **270**, 484.
7. Lipshutz, R. J., Fodpr, S. P. A., Gingeras, T. R., and Lockhart, D. J. (1999). *Nature Genet.*, **21**, 20.
8. Schena, M., Shalon, D., Davis, R. W., and Brown, P. O. (1995). *Science*, **270**, 467.

9. Carulli, J. P., Artinger, M., Swain, P. M., Root, C. D., Chee, L., Tulig, C., *et al.* (1998). *J. Cell. Biochem. Suppl.*, **30–31**, 286.

10. Gu, J., Guan, X. Y., and Ashlock, M. A. (1999). *Genome Res.*, **9**, 182.

11. O'Neil, M. J. and Sinclair, A. H. (1997). *Nucleic Acids Res.*, **25**, 2681.

12. Bertram, J., Palfner, K., Hidemann, W., and Kneba, M. (1998). *Anticancer Drugs*, **9**, 311.

13. Nezu, J., Oku, A., Jones, M. H., and Shimane, M. (1997). *Genomics*, **45**, 327.

14. DeRisi, J. L., Iyer, V. R., and Brown, P. O. (1997). *Science*, **278**, 680.

15. Martin, M. J., DeRisi, J. L., Bennett, H. A., Iyer, V. R., Meyer, M. R., Roberts, C. J., *et al.* (1998). *Nature Med.*, **4**, 1293.

16. Spellman, P. T., Sherlock, G., Zhang, M. Q., Iyer, V. R., Anders, K., Eisen, M. B., *et al.* (1998). *Mol. Biol. Cell*, **9**, 3273.

17. Meier-Ewert, S., Lange, J., Gerst, H., Herwig, R., Schmitt, A., Freund, J., *et al.* (1998). *Nucleic Acids Res.*, **26**, 2216.

18. Bishop, J. O., Morton, J. G., Rosbash, M., and Richardson, M. (1974). *Nature*, **250**, 199.

19. McCarthy, B. J. and Church, R. B. (1970). *Annu. Rev. Biochem.*, **39**, 131.

20. Zhao, N., Hashida, H., Takahashi, N., Misumi, Y., and Sakaki, Y. (1995). *Gene*, **156**, 207.

21. Nguyen, C., Rocha, D., Granjead, S., Baldit, M., Bernard, K., Naquet, P., *et al.* (1995). *Genomics*, **29**, 207.

22. Ross, M. T., Hoheisel, J. D., Monaco, A. P., Larin, Z., Zehetner, G., and Lehrach, H. (1992). In *Techniques for the analysis of complex genomes* (ed. R. Anune), p. 137. Academic Press, London.

23. Shalon, D., Smith, S. J., and Brown, P. O. (1996). *Genome Res.*, **6**, 639.

24. Bowtell, D. D. (1999). *Nature Genet.*, **21**, 25.

25. Marshall, A. and Hodgson, J. (1998). *Nature Biotechnol.*, **16**, 27.

26. Castellino, A. M. (1997). *Genome Res.*, **7**, 943.

27. Blattner, F. R., Plunkett, G. 3rd, Bloch, C. A., Perna, N. T., Burland, V., Riley, M. (1997). *Science*, **277**, 1453.

28. Tomb, J. F., White, O., Kerlavage, A. R., Clayton, R. A., Sutton, G. G., Fleischmann, R. D. (1997). *Nature*, **388**, 539.

29. Goffeau, A., Aert, R., Agostini-Carbone, M. L., Ahmed, A., Aigle, M., Alberghina, L. (1997). *Nature*, **387** (suppl), 5.

30. The *C. elegans* sequencing consortium. (1998). *Science*, **282**, 2012.

31. Persidis, A. (1998). *Biotechnology*, **16**, 393.

Chapter 11

Subtracted differential display and antisense approaches to analysing brain signalling

G. R. UHL and X. B. WANG

NIH, NIDA, Molecular Neurobiology Branch, 5500 Nathan Shock Drive, Baltimore, MD 21224, U.S.A.

Department of Neurology, Johns Hopkins School of Medicine, 725 North Wolfe Street, Baltimore, MD 21205-2185, U.S.A.

1 Introduction: rationale for differential display approaches when a little, some, or a lot of the target genome is understood

Many biological, pharmacological, and pathological processes can be better understood by defining the constellation of genes whose expression is altered by the process under study. The brain expresses more genes than most other tissues. Much of the richness of brain gene expression is reflected in variation in gene expression patterns from one brain region to the next, from one developmental stage to the next, from one drug treatment to the next, or from normal brain to diseased brain. Neuroscientists are thus particularly interested in applications of methods to help them identify and validate the significance of brain region-specific alterations in gene expression patterns.

A number of approaches to defining gene expression differences have been developed over the last several decades. Some elucidate changed expression of known genes and some allow identification of new genes based on their patterns of changed expression.

In this chapter, we initially relate our experience with a combination of approaches that have allowed us to clone novel drug-regulated genes in selected brain regions and to at least tentatively assign a biological role to a specific pattern of gene up-regulation. The initial approach used here, a combination of subtractive hybridization, PCR differential display (1), and use of antisense approaches for validation, provides a window on the general problem of identifying differential and/or varying gene expression in the brain (2).

Many such approaches were initiated when only a small fraction, *c.* 5%, of the mammalian genome was elucidated. As more and more of the genomes of several species of biological interest are elucidated, methods that take this knowledge into account become more powerful. We discuss the data from

207

differential display approaches in the context of more recent studies using methods such as microarray hybridization that allow more ready identification of changes in the expression of known genes. The data obtained here using a differential display variant provide novel species-specific gene sequences and illustrate the power of differential display approaches for use in species in which not all genes have been systematically elucidated.

1.1 Development of subtracted differential display (SDD)

Polymerase chain reaction-differential display (DD) was initially termed mRNA differential display when first reported in the early 1990s (3). This approach has been widely applied to identify the genes whose expression patterns are altered in different situations (4, 5). Several investigators, however, found relatively large numbers of false positive cDNAs with this procedure. In cDNA prepared from tissues with high mRNA complexity, many of the mRNAs that can be amplified through DD show apparently differential patterns of expression. However, RT-PCR or Northern blotting approaches are unable to confirm the differential expression in many cases. These false positives can be explained by variation at several experimental steps, including all of the vagaries of mRNA purification and preservation. The initial reverse transcription reaction at the beginning of differential display could provide cDNA differences despite similar mRNA abundance if the reverse transcriptase converts different mRNAs to cDNAs with different speeds or efficacies even under the same experimental condition. cDNAs of different lengths could share annealing sites for the arbitrary primers used for DD. A long cDNA may display multiple annealing sites, while a short one may have fewer. To reduce false positive DD results, we developed an approach called subtracted differential display (SDD) (1). The advantage of this approach is that it reduces the false positives by removing some of the unregulated cDNA using mRNA subtraction prior to DD. We have applied this approach to study gene regulation in brain, successfully identified a group of drug-regulated genes, and used antisense confirmation to identify the biological significance of up-regulation (2).

SDD consists of three steps:

(a) Reverse transcription of mRNA to yield single-stranded cDNA.
(b) Subtraction of cDNA from 'treated' samples using mRNA from control samples, and visa versa.
(c) PCR amplification and differential display of the subtracted cDNAs.

The following protocols were developed to identify influences of psychotropic drugs on gene expression in interesting brain regions of drug- or saline-treated rodents. This approach can also be used to identify genes whose expression patterns are altered in other circumstances. High quality mRNA and near-quantitative extraction of mRNA of this quality is of great importance for the entire experiment. Many companies provide mRNA extraction kits using oligo(dT) beads. The yield and the quality of mRNA yield may vary with the regents supplied in kits from different suppliers. Here, we describe a method using Fasttrack Kits to extract mRNA from brain specimens.

Protocol 1

mRNA preparation

Equipment and reagents

- Fasttrack Kits (Invitrogen)
- Dounce homogenizer
- Male Sprague-Dawley rats (250–300 g)
- Saline
- 7.5 mg/kg amphetamine

- Lysis buffer: 200 mM NaCl, 200 mM Tris–HCl pH 7.5, 1.5 mM $MgCl_2$, 2% SDS, and protein/RNase degrading enzymes
- Oligo(dT) cellulose beads

Method

1. Male Sprague-Dawley rats (250–300 g) were given intraperitoneal injections of 0.2 ml saline or amphetamine (7.5 mg/kg).

2. Brains were rapidly removed and regions dissected 4 h after injection. The dissected tissues were subject to mRNA extraction (6). The dissected tissue also can be frozen at −70°C for at least two months, although tissue frozen in −70°C freezers for more than 12 months often produces mRNA of lower quality.

3. Fresh or frozen brain tissue was homogenized with Dounce homogenizer in lysis buffer. The lysate was passed through a 21 gauge needle three times, then incubated at 45°C for 1 h. After adjusting the NaCl concentration to 0.5 M, oligo(dT) cellulose beads were added to the lysate at ratio of 1 mg per milligram of tissue. Tubes were rocked for 2 h at room temperature, the beads pelleted by centrifugation at 700 g for 5 min, and washed three times with binding buffer containing 500 mM NaCl, 10 mM Tris–HCl pH 7.5.

4. Oligo(dT)–mRNA beads were resuspended in 10 ml binding buffer. 1 ml oligo(dT)–mRNA bead suspension was applied to spin columns and spun at 700 g, 4°C for 5 min. The columns were washed three times with low salt buffer containing 250 mM NaCl, 10 mM Tris–HCl pH 7.5, and placed in new tubes. The beads were resuspended in100 μl DEPC treated water for 2 min, and then spun at 700 g for 5 min. Eluted mRNA was then ready for cDNA synthesis (*Protocol 2*).

5. The remaining 9 ml of oligo(dT)–mRNA beads were centrifuged at 700 g at 4°C for 5 min and resuspended in 1 ml binding buffer. Beads were stored on ice or at 4°C for cDNA subtraction (*Protocol 3*).

100 μl mRNA eluted from 1 ml oligo(dT)–mRNA (*Protocol 1*, step 4) was used as template for synthesis of single-stranded cDNA.

Initial experiments suggested that the above cDNA to mRNA ratio, 1:10, was optimal for experiments in which DD was subsequently performed. The ratio between cDNA and mRNA can be varied to produce subtractions of differing efficiencies, and can be conveniently manipulated by reducing the cDNA volume. Using 100 μl or 10 μl cDNA to hybridize with 1 ml mRNA beads in the above protocol will cause the subtraction ratio to vary from 1/10 to 1/100. Monitoring the efficiency of extraction of radiolabelled cDNA can help determine the subtraction efficacy. In initial experiments, we found that fivefold mRNA excess could remove about 40% of the cDNA. Tenfold mRNA excess removed more than 80% of the cDNA. Increases in this ratio as high as 20-fold excess did not appreciably increase the fractional cDNA removal under the hybridization conditions used here.

Protocol 2

cDNA synthesis

Reagents

- Oligo(dT) primer
- 50 mM Tris–HCl buffer pH 8.3 at 42 °C
- dNTPs
- RNase inhibitor
- KCl
- MgCl$_2$
- Reverse transcriptase

Method

1. 100 µl mRNA was aliquoted into five tubes each containing 20 µl. 5 µl of oligo(dT) primer yielded a final concentration of 20 pmol. Tubes were heated to 70°C for 2 min and then quenched on ice for 2 min.

2. First strand cDNA synthesis was primed in a total volume of 50 µl reaction solution containing 50 mM Tris–HCl buffer pH 8.3 at 42°C, 0.5 mM dNTPs, 1 U/µl RNase inhibitor, 75 mM KCl, 3 mM MgCl$_2$, and 250 U reverse transcriptase.

3. 1 h later the reaction was terminated by adding 8 µl of 6 M NaOH. The reaction mix was then incubated at 60°C for 30 min to digest RNA. After RNA digestion, 8 µl of 6 M HCl was added to neutralize the NaOH.

4. cDNA solutions from each of the five tubes were pooled for a total volume of 250 µl prior to mRNA subtraction procedures.

Protocol 3

mRNA subtraction

Reagents

- 10 mM Tris–HCl pH 7.4
- Sodium chloride
- Poly(A) (18-mer)

Method

1. 250 µl cDNA prepared from drug-treated brains (*Protocol 2*) was heated to 100°C for 2 min and cooled on ice for 2 min.

2. 250 µl of denatured cDNA was mixed with 1 ml mRNA-loaded oligo(dT) beads (*Protocol 1*, step 5) and hybridized at room temperature for at least 12 h in binding buffer containing 10 mM Tris–HCl pH 7.4, 500 mM sodium chloride, and 100 nM poly(A) (18-mer).

3. After hybridization, beads carrying the poly(A)$^+$ RNA–cDNA complex and unhybridized poly(A)$^+$ RNA were centrifuged 700 g r.p.m. for 5 min. 800 µl supernatant was carefully transferred, avoiding contamination by beads, to a new tube. This supernatant contained the unhybridized, subtracted cDNA. The supernatant was then dialysed in DEPC water for 6–18 h and concentrated to 100 µl by lyophilization.

Protocol 4

Differential display of subtracted cDNAs

Equipment and reagents

- Polyacrylamide sequencing gel electrophoresis equipment
- PCR reaction buffer: 10 mM Tris–HCl buffer (pH 8.3 at 25 °C), 50 mM KCl, 1.5 mM MgCl$_2$, 0.001% gelatin, 0.2 mM dNTPs, 1 pM [^{35}S]dATP, 5 µM poly(dT) primer, 2.5 µM arbitrary primer, 2.5 U/100 µl AmpliTaq DNA polymerase
- X-ray film
- Anchor primer: 5′-TTTTTTTTTTTT(G/C)C-3′
- Arbitrary primer: 5′-GGACAGCTTC-3′
- PCRII vector (Invitrogen)
- Qiagen tips and resin (Qiagen Inc.)

Method

1. Subtracted cDNAs were amplified by differential display PCR using conditions similar to those of Liang and Pardee (3). The anchor primer and arbitrary primer was previously described (3); other sequences attempted in preliminary experiments did not work as well.

2. Polymerase chain reactions were performed in PCR reaction buffer. 40 thermal cycles of 94 °C for 30 sec, 40 °C for 2 min, and 72 °C for 1 min, were followed by a final extension at 72 °C for 5 min.

3. Amplified radiolabelled cDNAs were separated on 6% polyacrylamide sequencing gels by electrophoresis at 80 V for 3 h. Gels were blotted, dried, and subjected to film autoradiography.

4. cDNA bands showing substantial intensity differences between the saline- and drug-treated striatum were excised.

5. Excised gel slices were rinsed twice for 5 min with water, and cDNA was eluted by boiling gel slices in 50 µl water for 5 min.

6. 5 µl of the recovered cDNA was reamplified in a 50 µl reaction volume using the same primer sets and thermal cycling parameters noted above, omitting radiolabelled precursors.

7. 5 µl of each reamplified PCR product was subcloned into the PCRII vector.

8. Plasmid DNA was extracted and purified using Qiagen tips and resin, and subjected to sequencing using manual and automated methods (Applied Biosystems, Inc.). Sequences were compared using *BLAST* (7).

2 Example: candidate drug-regulated genes from striatum

Using SDD, we have identified several candidate genes apparently regulated by psychotropic drugs. As shown in *Figure 1*, apparent changes in expression patterns include up- and down-regulation after drug treatment. Expression changes in about 70% of the candidate genes were reconfirmed by RNA analysis using Northern blot and/or RNase protection assays. At least one-third thus still represented false positive results of the SDD procedure. Sequencing results provided information about these drug-regulated genes. Based on the sequences, we focused on several interesting genes including a novel G protein beta$_1$ subunit and calcineurin, a neuron-selective phosphatase (1, 2).

3 Antisense confirmation: one practical approach

Identification of drug-induced changes in gene expression led us to question their *in vivo* relevance. To elucidate possible relevance of drug-induced gene expression changes to long-term consequences of drug action, we examined the effects of transient attenuation of the expression of one drug regulated gene, rGβ_1, on a long-term behavioural adaptation to the psychostimulant treatments that induced rGβ_1 up-regulation. To reduce rGβ_1 protein production, we applied a microinjection technique to introduce a specific anti-rGβ_1 oligonucleotide into the cerebrospinal fluid perfusing structures including the nucleus accumbens. The accumbens is the brain region in which we found some of the most prominent drug-induced changes in neuronal expression of rGβ_1 immunoreactivity. The accumbens plays important roles in drug-induced behavioural changes.

3.1 Antisense design

An anti-rGβ_1 oligonucleotide complementary to sequences flanking the ATG start codon of the rGβ_1 cDNA (5'-TTCACTCATCTTCACGTC) was designed to block the translation of rGβ_1. The sequence 5'-ATCGCTCGTCATCTCGTC was designed as a control missense oligonucleotide.

3.2 Antisense administration

300 g male Sprague-Dawley rats were anaesthetized with pentobarbital and implanted with 22 gauge guide cannulas with tips 1 mm dorsal to the left lateral cerebral ventricles. Oligonucleotides were injected on alternate days beginning

five days following guide cannula implantation. 2 μl of artificial CSF containing 10 μg $rG\beta_1$ antisense or missense were injected intracerebroventricularly using Hamilton syringes fitted through the guide cannulae. Neither the animals' gross appearance, gross behavioural activity, body weight, nor magnitude of locomotion responses to acute administration of amphetamine or cocaine were altered by these treatments.

3.3 Confirmation of $rG\beta_1$ protein expression level

$rG\beta_1$ expression was monitored by both Western blot analysis and immunohistochemistry using specific antibodies directed against $rG\beta_1$. The levels of $rG\beta_1$ immunoreactivity in striatum were reduced one day after second infusion of anti-$rG\beta_1$ oligonucleotide. This inhibition ranged up to 70% after the fourth anti-$rG\beta_1$ oligonucleotide infusion. A missense oligonucleotide infused on the same schedule had no effect (*Figure 1*). Such documentation is vitally important. Other oligonucleotide sequences used for these experiments were ineffective in reducing levels of $rG\beta_1$ protein, and were discarded for behavioural experiments. While using oligonucleotides directed close to translational start sequences may provide a greater chance of success, no clear-cut rules now definitively separate

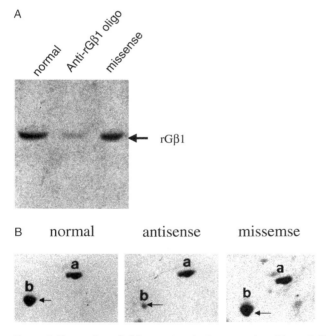

Figure 1 Expression of $rG\beta_1$ gene products was reduced by anti-$rG\beta_1$ oligonucleotide treatments. Rats were given four intracerebroventricular infusions of anti-$rG\beta_1$ oligonucleotide or missense control nucleotides. The expression of $rG\beta_1$ gene product in rat striatum-accumbens was detected by one-dimensional (A) or two-dimensional (B) Western blotting using antisera that detect a stable, non-regulated band ('a') and authentic $rG\beta_1$ which migrates at the position of 'b'.

sequences that are successful in reducing expression levels from those that do not. Trial and error can still be involved in antisense experiments.

3.4 Evaluation of antisense effects on drug treated rats: effects on locomotor 'sensitization'

Daily intraperitoneal injections of 20 mg/kg cocaine were begun six hours after the fourth oligonucleotide injection and continued for five days. Locomotor responses to the first and fifth cocaine injections were analysed by photocell beam breaks. Rats were habituated to the apparatus for 50 minutes, injected with cocaine, and monitored for 150 minutes to assess locomotion. Anti-rGβ_1 DNA-injected rats displayed patterns of enhanced locomotion after initial injection of 20 mg/kg cocaine that were indistinguishable from those noted in control animals injected with missense control DNAs. In the rats injected with missense DNA, marked behavioural sensitization was also noted after the five daily cocaine administrations. Locomotor scores in these animals doubled those induced by the cocaine dose that preceded sensitization. By contrast, however, animals treated with anti-rGβ_1 oligonucleotides revealed no evidence for behavioural sensitization. This effect of anti-rGβ_1 oligonucleotide treatments was reversible. Each of the three groups of anti-rGβ_1 pre-treated rats displayed full recovery of cocaine-induced sensitization when new sensitizing treatments were begun three weeks after cessation of antisense treatments. When antisense treatments were administered following establishment of sensitization but before its expression in two additional groups of animals, the antisense treatments failed to alter the robust expression of the behavioural sensitization in pre-sensitized rats. These results provide substantial evidence that rGβ_1 or a very closely related protein is required for establishment of cocaine-induced behavioural sensitization.

4 Pitfalls

False positive and false negative results can each come from these sorts or approaches. Variability in quantitative yields from mRNA extractions and cDNA preparation can result from individual variability in tissue and RNA handling. Parallel preparation of multiple samples from drug-treated and control animals provides one invaluable method for assessing the degree of this variability, so that the extent of 'drug' versus 'placebo' differences can be compared and contrasted to the 'noise' from sample to sample using these exacting techniques. It is important to provide early confirmation that levels of expression of mRNAs that correspond to cDNAs whose intensities appear to be different in differential display gels are in fact different. Use of other techniques, such as RNase protection or Northern analyses, is key to this process. It is important that the RNA source for the Northern and or RNase protection analyses be different from that used for the differential display, so that artefactually-induced differences in mRNA based on its extraction, for example, are not falsely ascribed to differences in gene expression. False positive results can also arise from other sources, including the possibility that several cDNAs can be subcloned from DNA amplified from DNA extracted from a single excised band from a differential display gel.

False negative results could come from sources similar to those that yield false positive data. Low abundance mRNAs can readily be amplified in the differential display assays despite copy numbers so low that they may be difficult to detect in Northern analyses. In a complex tissue such as the brain, there can be large regional differences in levels of mRNA expression and regulation. True differences in small brain regions can be masked if too large regions are used for mRNA preparation. This can be especially difficult if other cells in the larger tissue chunk sampled do not alter expression, or alter it in opposite directions.

5 Comparisons with other methods

A number of classical methods have been successful in identifying differentially expressed genes. Differential screening of cDNA libraries using hybridization probes made from different mRNA sources provides one such approach. Subtracting cDNA with mRNA from another source and using this subtracted cDNA to screen libraries provides another method. However, many such approaches depend critically on the construction of high quality cDNA libraries. These are technically difficult, time-consuming, and costly steps. Such approaches do provide an antidote to one limitation of DD and PCR-DD approaches, which usually require re-screening cDNA libraries after DD. This is especially true if full-length cDNAs are required, since the vast majority of DD products are typically short 3' non-coding region gene fragments.

As shown in *Table 1*, a critical disadvantage of mRNA differential display has often been low reliability. This problem has concerned many investigators. As noted above, in our preliminary experiments more than half of the apparently differentially expressed bands produced by DD were false positives. These false positive PCR products were significantly reduced by the mRNA–cDNA subtraction technique (*Figure 2*). Thus, after cDNA subtraction, the reliability of DD

Table 1 A comparison of the cDNA library screening with DD

	Subtractive cDNA library screening	mRNA differential display
Key technique	Need construction of subtracted library	RT-PCR
Sensitivity	Difficult to identify low abundance genes	Sensitive
Reliability	Possibly more reliable	False positives better documented
Experiment cycle	Longer	Shorter
Expense	Higher	Lower
Re-screening library	Possibly not	Needed
Amount of RNA for experiment	100–500 µg	2–5 µg

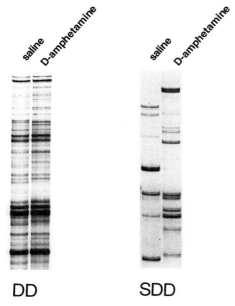

Figure 2 Comparison of differential display (DD, *left*) with subtracted differential display (SDD, *right*) PCR. cDNA from hippocampi of rats treated with saline or D-amphetamine was amplified by differential display PCR before (DD) or after (SDD) subtraction with tenfold excess mRNA from hippocampi of saline-treated rats.

was increased more than 80% using Northern and RT-PCR analyses as the 'gold standards' for comparison (1). Recent modifications to the DD approach, including removal of DNA contamination of the mRNA using DNase (7) and employment of longer PCR primers (8), may also help to improve the reliability and reproducibility of the SDD variant of the basic DD approach.

6 Conclusions

As a PCR-based screening method, SDD is an advanced approach that can be applied to identify differentially expressed genes even in high mRNA complexity tissues such as the brain. A number of other techniques including RNA fingerprinting (9), representational difference analysis (RDA) (10), serial analysis of gene expression (SAGE) (11), and micro-gene arrays or DNA chips have also been applied to identify differentially expressed genes. These differentially expressed genes, either known or unknown, can provide important clues for studying and characterizing biochemical functions and pharmacological processes. Elucidation of differentially expressed genes will add to understanding mechanisms and therapeutic approaches to many human diseases, including those that influence the nervous system.

References

1. Wang, X. B. and Uhl, R. G. (1998). *Mol. Brain Res.*, **53**, 344.
2. Wang, X. B., Funada, M., Imai, Y., Revay, R. S., Ujike, H., Vandenbergh, D. J., *et al.* (1997). *J. Neurosci.*, **17**, 5993.
3. Liang, P. and Pardee, A. B. (1992). *Science*, **257**, 961.
4. Wang, X. (1996). *Trends Pharmacol. Sci.*, **17**, 276.
5. Livesey, F. J. and Hunt, S. P. (1996). *Trends Neurosci.*, **19**, 84.
6. Simada, S., Kitayama, S., Lin, C.-L., Nanthankumar, E., Gregor, P., Patel, A., *et al.* (1991). *Science*, **254**, 576.
7. Liang, P., Averboukh, L., and Pardee, A. B. (1993). *Nucleic Acids Res.*, **21**, 3269.
8. Zhao, S., Ooi, S. L., and Pardee, A. B. (1995). *BioTechniques*, **18**, 842.
9. Welsh, J., Chada, K., Dalal, S., Cheng, R., Ralph, D. and McClelland, M. (1992). *Nucleic Acids Res.*, **20**, 4965.
10. Lisitsyn, N., Lisitsyn, N., and Wigler, M. (1993). *Science*, **259**, 946.
11. Velculescu, V. E., Zhang, L., Vogelstein, B., and Kinzler, K. W. (1995). *Science*, **270**, 484.

Chapter 12

Differential display analysis of memory-associated genes

A-MIN HUANG* and EMINY H. Y. LEE[†]
*Department of Physiology, National Cheng-Kung University Medical College, Tainan 701, Taiwan, Republic of China
[†]Institute of Biomedical Sciences, Academia Sinica, Taipei 115, Taiwan, Republic of China

1 Introduction

Memory formation is a complicated process, especially in mammals. In the form of long-term memory formation, it is well known that it requires *de novo* RNA and protein synthesis because protein and mRNA synthesis inhibitors impair long-term memory formation (1, 2). In the invertebrate system, extensive efforts have been made to identify specific genes associated with memory formation. For example, by using quantitative 2D gel analysis, Castellucci *et al.* (3) have identified several proteins that are related to the process of long-term sensitization, a form of long-term memory studied in the invertebrate *Aplysia*. Also, screening of *Drosophila* mutants has yielded about ten genes that are related to the process of olfactory learning and memory (4). Although these methods are effective, two disadvantages exist by using these approaches for studies in the mammalian system. First, they take a long time to identify and characterize the genes of interest. Secondly, gene expression associated with learning and memory in mammals is much more complicated which may not be well detected by using the traditional methods.

PCR differential display is a powerful method which can be used to identify specific and rare genes in a tissue or cell system (see Chapter 1). Liang and Pardee (5) first used this method to identify genes differentially expressed in a normal versus a human breast cancer cell line. We later adopted the same method to examine gene expressions which are specifically associated with memory formation of one-way inhibitory avoidance learning in rats. By using this method, we have identified at least two genes which are specifically related to memory formation in rats. First, we have found that the expression of integrin-associated protein (IAP) gene is approximately fourfold higher in rats showing good retention performance than rats showing poor retention performance. This enhanced IAP gene expression is confirmed by both *in situ* hybridization and quantitative RT-PCR analysis. It is further confirmed by antisense oligonucleotide manipulation in long-term potentiation (LTP) study *in vivo* (6), a synaptic model for learning and memory (7). Glial fibrillary acidic protein (GFAP) is another gene that has been identified which is differentially expressed between good memory and poor memory rats. However, the expression of this gene turns out to belong to DNA polymorphism, in that the good memory rats preferentially expressed one isotype of the GFAP gene, but the poor memory and control rats preferentially expressed the other isotype of the same gene (8). The details of these approaches are described below and the role and mechanism of these genes involved in memory consolidation are currently under investigation.

2 Behavioural training and testing

Because the learning and memory processes in mammals are much more complicated than that in the invertebrate, it would be more difficult to identify specific genes associated with a complicated form of learning in mammals. Therefore, we have, at the beginning, adopted a simple form of learning task that allows us to apply the differential display screening method more easily. We adopt the one-way inhibitory avoidance learning task as the behavioural paradigm (*Protocol 1*).

Protocol 1

One-way inhibitory avoidance learning task

Equipment and reagents

- Sprague-Dawley rats, eight weeks old
- A behavioural apparatus consisting of a trough-shaped alley divided by a sliding door into an illuminated safe compartment and a dark compartment
- A shock generator with facilities to vary current connected to the floor of the dark compartment
- A lamp with 20 candle-light in intensity
- A timer

Behavioural procedures: conduct the training and testing procedures between 10:00 a.m. and 5:00 p.m.

A. Training

1. Keep rats in a dim room for 1 h to adjust the environment.
2. On the training phase, put the rat at the far end of the illuminated compartment facing away from the door.
3. As the rat turns around, open the door.
4. When the four paws of the rat step into the dark compartment, close the door and give a 1.0 mA/1 sec foot shock.
5. Remove the rat from the alley and put it back to its home cage.

B. Testing

1. 3 h or 24 h later, conduct the retention test.
2. Put the rat again in the illuminated compartment and start the timer. Stop the timer when the four paws of the rats step into the dark compartment. The time interval is the retention latency measure.

3. Rats which do not enter the dark compartment and reach the ceiling score of 600 sec are removed from the alley and assigned as good memory rats. Rats with a retention latency less than 80 sec are assigned as the poor memory rats. The remaining rats showing a retention latency between 80 and 600 sec are not used in the present study.

4. The untrained rats and rats received electric shock only without going through the training procedure are used as the control.

3 Hippocampal tissue dissection and total RNA isolation

Hippocampus is a brain structure which is believed to be involved in memory formation (9). In our laboratory, we have found that blocking the synthesis of mRNAs or proteins in the hippocampus impairs memory retention in rats (2). For the analysis of genes associated with memory formation, we first chose the hippocampus as the target tissue. Total hippocampal RNA isolated from poor memory rats and good memory rats are then compared by the PCR differential display method. Total RNAs are purified with an improved one-step extraction method using the Ultraspec II RNA isolation system (Biotecx Laboratories), by which multiple samples can be isolated simultaneously (*Protocol 2*). Because trace amounts of chromosomal DNA contamination in the RNA samples could be amplified along with mRNAs, removal of all contaminating DNA is carried out routinely after total RNA isolation (*Protocol 3*).

Protocol 2

Tissue dissection and total RNA purification

Equipment and reagents

- Hand-held glass homogenizer
- Ultraspec II RNA isolation system (Biotecx Laboratories)
- Ice-cold saline
- Chloroform
- Isopropanol
- DEPC treated H_2O

Method

1. Immediately after the memory retention test, kill the rat by decapitation, remove the brain, and place it in ice-cold saline for 5 min.

2. Dissect out the hippocampal tissues on an ice-cold platform using a slicer. Freeze the tissues on dry ice immediately and store at $-80\,°C$ until use.

3. Homogenize the hippocampal tissues (80–100 mg) extensively (20–25 strokes) in a hand-held pre-cooled 1 ml glass homogenizer containing 1 ml (1 ml/10–100 mg tissue) of the ice-cold Ultraspec RNA isolation solution. Store the homogenate on ice for 5 min.

Protocol 2 continued

4. Transfer the homogenate to autoclaved tubes, add 0.2 ml of chloroform, shake vigorously for 15 sec, and keep on ice for 5 min.

5. Centrifuge the homogenate at 12 000 r.p.m. at 4°C for 15 min. Transfer the upper aqueous layer (4/5 volume) to a clean tube.

6. Precipitate RNA by addition of 0.5 vol. of isopropanol, mix well, and leave on ice for 30 min.

7. Centrifuge as in step 5. Discard the supernatant. Wash the pellet twice with 1 ml of 75% EtOH (made with DEPC H_2O).

8. Discard the supernatant, briefly re-spin the tube, and remove any traces of ethanol using the pipette tip.

9. Dry the pellet briefly for a few minutes in vacuum.

10. Resuspend the pellet in 100 μl of DEPC treated water.

11. Read the optical density of OD_{260} and OD_{280} of the RNA sample to estimate the quantity and purity of the total RNA.

Protocol 3

Removal of DNA contamination

Reagents

- DNase I (RNase-free)
- 10 mM Tris–HCl pH 9.0, 50 mM KCl, 1.5 mM $MaCl_2$, and 0.1% Triton X-100
- Phenol:chloroform (3:1)

- 2.5 M potassium acetate
- Absolute ethanol

Method

1. Incubate 20 μg of total RNA with 2 U of DNase I (RNase-free) in 10 mM Tris–HCl pH 9.0, 50 mM KCl, 1.5 mM $MaCl_2$, and 0.1% Triton X-100 at 37°C for 30 min.

2. Stop the reaction of DNase I by adding an equal volume of phenol:chloroform (3:1) to the sample.

3. Mix by vortexing and centrifuge the sample at 12 000 g for 10 min at 4°C.

4. Transfer the supernatant to a clean tube and precipitate the RNA by adding 2 vol. of absolute EtOH and 0.1 vol. of 2.5 M potassium acetate, mix well, and incubate at −80°C for 30 min.

5. Pellet and wash the RNA as in *Protocol 2*, steps 7–9.

6. Redissolve the RNA in 20 μl of DEPC H_2O.

7. Read the optical density of OD_{260} and OD_{280} of the RNA sample to estimate the quantity and purity of the total RNA.

8. Store the RNA sample at −80°C.

4 PCR differential display analysis

The PCR differential display method adopted in our research is based on the first and second generations of this method described by Liang *et al.* (5, 10). For the first generation PCR differential display method, the two bases-anchored oligo(dT) primers are $T_{12}MN$. M = A, C, G, or T; N = A, C, or G. 12 RT reactions are needed for each RNA sample. For the second generation PCR differential display, the two bases-anchored oligo(dT) primers are $T_{12}VN$. V is a mixture of A, C, G, and T; N = A, C, or G. Only four RT reactions are needed for each RNA sample. The arbitrary primers for both of these two methods are 10-mer in random sequence. 30 arbitrary 10-mer primers are used in our study (*Table 1*). The GC content of these primers range from 40–70%. For the third generation PCR differential display, the anchored oligo(dT) primers are three 16-mers, each with a *Hind*III restriction site, a T_{11}, stretch and a single 3' base, either A, C, or G (designated as $HT_{11}A$, $HT_{11}C$, and $HT_{11}G$). Arbitrary primers are a series of 13-mers with a *Hind*III restriction site (11, 12). Primers and reagents for the second and the third generation of PCR differential display analyses are available from GeneHunter Company (RNAimage kits). The principle methods include reverse transcription, PCR amplification, and sequencing gel analysis. Here we show an example for the differential display analysis of six RNA samples: three from poor memory rats and three from good memory rats.

Table 1 Sequences of arbitrary primers used in this study (5'→3')

CTGATCCATG	CAAACGTCGG	GAGTGTCTCG
CTGCTCTCAG	GTTGCGATCC	CACATAGCGC
CTTGATTGCC	CTAGGTCTGC	CGAAGCGATC
CGAAACAGTC	AGGGGTCTTG	CCCTCATCAC
CGCTGTTACC	CAAAACAGCA	CCTGTTAGCC
CTGCATAGGT	CCATTGAAGG	GCAGCTCATG
AGGTCACTGA	CAAGAGCCAG	CGCTTGCTAG
AGCCAGCGAA	ATTGGCTGGC	GAACCTACGG
GACCGCTTGT	CACACGCACC	CTAGCTGAGC
AGGTGACCGT	GAAAGTGGAC	GAGGAGGCTG

4.1 Reverse transcription

Total RNAs are reverse transcribed in this reaction (*Protocol 4*). In the standard protocol, duplicate identical samples from each RNA preparation are run side by side (10, 13). In our laboratory, duplicate or triplicate RNA samples from different individual animals are used for comparison simultaneously. For example, we use RNA samples from three poor memory rats and three good memory rats.

Protocol 4

Reverse transcription

Reagents

- Anchored oligo(dT) primer (25 μM)
- Superscript II MMLV (Gibco BRL)
- 5 × RT buffer: 250 mM Tris–HCl pH 8.3, 375 mM KCl, 15 mM MgCl$_2$ (comes with Superscript II MMLV)
- 0.1 M DTT (comes with Superscript II MMLV)
- 200 μM dNTP mix
- RNasin (40 U/μl)

Method

1. In six autoclaved ice-cold microcentrifuge tubes add 1 μg of each total RNA sample to each tube, 2 μl of one of anchored oligo(dT) primer (25 μM), and DEPC H$_2$O to a final volume of 9 μl.

2. Incubate at 70°C for 10 min.

3. Quench the tubes on ice for 5 min.

4. Spin the tubes briefly to collect condensation and set tubes on ice.

5. Make a core mix for the RT reaction (6.5 × core mix for six RNA samples):

	1 ×	6.5 × core mix
5 × RT buffer	4 μl	26 μl
200 μM dNTP mix	1 μl	6.5 μl
0.1 M dithiothreitol (DTT)	2 μl	13 μl
RNasin (40 U/μl)	0.28 μl	1.82 μl
Superscript II MMLV (200 U/μl)	1 μl	6.5 μl
DEPC H$_2$O	2.72 μl	17.68 μl

6. Add 11 μl of the core mix to each RNA sample. Mix by gentle pipetting.

7. Incubate at 37°C for 1 h.

8. Terminate the reaction by heating the sample at 80°C for 10 min. Place the tubes on ice. Spin the tubes briefly to collect condensation.

9. Set tubes on ice for PCR, or store at −80°C for later use.

4.2 PCR amplification

Different combinations of primer set are used for the PCR amplification (*Protocol 5*). Four anchored primers in combination with 30 arbitrary primers give a total of 120 sets of primer pairs. For the six RT reactions, 720 PCR reactions have been carried out. In our laboratory, 30 PCR reactions can be analysed on a sequencing gel which has 32 sample wells (*Protocol 6*). A total of 24 sequencing gels are needed to analyse these 720 PCR reactions.

Protocol 5

PCR amplification

Reagents

- 10 × PCR buffer: 100 mM Tris–HCl pH 8.8, 500 mM KCl, 1% Triton X-100 (this is the buffer without MgCl$_2$ from Promega)
- 25 mM MgCl$_2$
- 40 μM dNTP
- [α-^{35}S]dATP (> 1000 Ci/mmol, 10 mCi/ml)
- Taq DNA polymerase

Method

1. Make a core mix for the PCR reactions (33 × core mix for 30 PCR reactions):

	1 ×	33 × core mix
• Sterile H$_2$O	4 μl	132 μl
• 10 × PCR buffer	2 μl	66 μl
• 25 mM MgCl$_2$	0.96 μl	31.68 μl
• 40 μM dNTP	2 μl	66 μl
• 5 μM arbitrary 10-mer	2 μl	66 μl
• 25 μM one of the T$_{12}$VN primers	2 μl	66 μl
• [α-^{35}S]dATP (> 1000 Ci/mmol, 10 mCi/ml)	0.5 μl	16.5 μl
• Taq DNA polymerase	0.2 μl	6.6 μl

2. Aliquot 18 μl of the core mix into each 0.5 ml microcentrifuge tube.

3. Add 2 μl of each RT products to the 18 μl of the PCR reaction mixture. Mix gently by pipetting up and down.

4. Overlay 30 μl mineral oil in each tube.

5. Run PCR as follows: 40 cycles of 94 °C for 30 sec, 40 °C for 2 min, 72 °C for 30 sec. Then 72 °C for 5 min, and 4 °C for soaking.

4.3 Sequencing gel analysis

Protocol 6

Sequencing gel analysis

Equipment and reagents

- DNA sequencing gel apparatus
- X-ray film
- 1 × TBE buffer: 0.089 M Tris base, 0.089 M boric acid, and 0.002 M EDTA
- Urea
- 40% acrylamide: bis (19:1) solution
- Loading dye

Protocol 6 continued

Method

1. Prepare a 6% denaturing polyacrylamide gel in $1 \times$ TBE buffer. In a 250 ml beaker, add:

   - 28.8 g urea
   - 27.6 ml H_2O
   - 6 ml of $10 \times$ TBE buffer
   - 9 ml of 40% acrylamide: bis (19:1) solution
   - Final volume 60 ml

2. Stir mix until urea is dissolved. Let the gel polymerize for at least 2 h before using.

3. Pre-run the gel at 60 W constant power for 30 min.

4. Mix 9 μl of each sample with 3.6 μl of loading dye; incubate at 80 °C for 5 min immediately before loading onto a 6% DNA sequencing gel.

5. Run the gel at 60 W constant power (with voltage not to exceed 1800 V) in $1 \times$ TBE buffer for about 3 h until the xylene cyanol dye (the slower moving dye) is about 5 cm from the bottom. Turn off the power supply and blot the gel onto a piece of 3MM paper. Cover the gel with a plastic wrap and dry it at 80 °C for 1 h.

6. Orient the dried gel and an X-ray film with needle punches and expose at room temperature overnight.

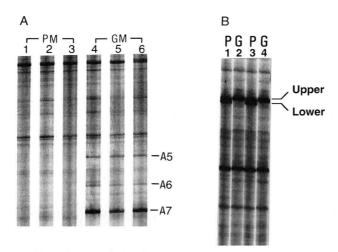

Figure 1 PCR differential display of rat hippocampal RNA associated with memory formation of inhibitory avoidance learning. (A) Total RNA samples isolated from the hippocampus of three poor memory (PM) and three good memory (GM) rats tested three hours after training were compared. The sequence for the 5′ arbitrary primer is 5′-AGC CAG CGA A-3′ and the 3′ anchored primer is $T_{12}VA$ (V is a mixture of A, C, and G). (From ref. 6.) (B) Total RNA samples isolated from two poor memory (P) and two good memory (G) rats tested 24 hours after training were compared. The sequence for the 5′ arbitrary primer is 5′-CTG ATC CAT G-3′ and the 3′ anchored primer is $T_{12}AG$. (Huang and Lee, unpublished results.)

Figure 1 shows representative differential display obtained by two combinations of primer pairs. At different times after training, some cDNAs are differentially expressed between samples from the poor memory and good memory rats. *Figure 1A* shows that three hours after training, three cDNA bands, designated as A5, A6, and A7, are expressed in higher level in the good memory rats when the anchored primer $T_{12}VA$ is paired with the arbitrary primer 5'-AGC CAG CGA A-3'. *Figure 1B* shows that 24 hours after training, an extra band, designated as the lower band, has appeared in the samples from the poor memory rats. In this case, the anchored primer $T_{12}AG$ is paired with the arbitrary primer 5'-CTG ATC CAT G-3'.

4.4 Recovery and reamplification of differentially expressed cDNA bands

The differentially expressed cDNA fragments can be recovered from the sequencing gel rapidly and reamplified with the same set of primers used for the PCR differential display analysis (10–12) (*Protocol 7*).

Protocol 7

Recovery and reamplification of differentially expressed cDNA bands

Reagents

- See *Protocol 5*
- 3 M NaOAc
- 10 mg/ml glycogen
- Ethanol

Method

1. After developing the film, orient the autoradiogram with the gel. Fix the film and the autoradiogram with stapes and put upon a light box.
2. Locate cDNA bands of interest by marking with a clean pencil on the 3MM paper.
3. Cut out the cDNA bands with a clean surgery blade.
4. Soak the gel slice along with the 3MM paper in 100 μl sterile H_2O for 10 min.
5. Boil the tube for 15 min.
6. Spin in a microcentrifuge for 10 min. Transfer the supernatant to a new microcentrifuge tube.
7. Add 10 μl of 3 M NaOAc, 5 μl glycogen (10 mg/ml), 2 vol. of absolute EtOH, and mix well.
8. Precipitate DNA at $-80°C$ for 30 min. Spin at $4°C$ for 10 min to pellet DNA.
9. Wash the DNA pellet with 200 μl ice-cold 75% EtOH.
10. Spin briefly and remove residual EtOH.
11. Dissolve the DNA pellet in 10 μl of sterile H_2O and use 4 μl for reamplification.

12. Reamplify the recovered cDNA with the same set of primer and PCR conditions except the dNTP concentrations are at 20 μM and no isotope is added. For a 40 μl reaction, combine the following components:

- 20.4 μl sterile H_2O
- 4 μl of 10 × PCR buffer
- 2 μl of 200 μM dNTP
- 4 μl of 2 μM arbitrary primer
- 4 μl of 2 μM one of the $T_{12}VN$ primer
- 4 μl cDNA templates from step 11
- 0.4 μl *Taq* DNA polymerase (5 U/μl)

13. Run 10 μl of the PCR products on a 1.5% agarose gel. Check to see if the size of the reamplified PCR products are consistent with their sizes on the denaturing poly-acrylamide gel.

14. Clone the freshly reamplified cDNA products by the TA vector.

5 Cloning and sequencing of the differentially expressed products

5.1 TA cloning

The differentially expressed cDNA fragments can be quickly cloned by using the TA cloning system. *Taq* polymerase adds a single deoxyadenosine (A) to the 3′ ends of the PCR products. The PCR products can be ligated to a plasmid vector which is linear and has single 3′ deoxythymidine (T) residues. The pGEM-T vector (Promega) and pCR2.1 vector (Invitrogen) have been used in our laboratory. Cloning procedures by using the pCR2.1 vector (*Protocol 8*) are described here.

Protocol 8

Cloning the reamplified PCR products by a TA vector pCR2.1

Reagents

- pCR2.1 vector
- T4 DNA ligase
- One Shot™ competent cells (Invitrogen)

- SOC medium: 2% tryptone, 0.5% yeast extract, 10 mM NaCl, 2.5 mM KCl, 10 mM $MgCl_2.6H_2O$, and 20 mM glucose (provided with the Invitrogen TA cloning kit)

A. Ligation

1. In a microcentrifuge tube, combine the following components:

- X μl fresh PCR product

- 1 μl of 10 × ligation buffer
- 2 μl pCR2.1 vector (25 ng/μl)
- (9 – X) μl sterile water
- 1 μl T4 DNA ligase (4.0 Weiss units)

$$X = \frac{(Y \text{ bp PCR product}) (50 \text{ ng pCR2.1 vector})}{(\text{size in bp of the pCR 2.1 vector: } \sim 3900)}$$

2. Incubate at 14°C overnight. Take out 2 μl of the ligation products for transformation. Store the remaining products at 4°C.

B. Transformation, selection, and screening

1. Spin the tubes containing the ligation products briefly and place them on ice.

2. Thaw on ice one 50 μl vial of frozen One Shot competent cells for each ligation reaction.

3. Pipette 2 μl of each ligation reaction directly into the competent cells and mix by stirring gently with the pipette tip. Incubate the vials on ice for 30 min.

4. Heat shock for exactly 30 sec in the 42°C water-bath. Do not mix or shake.

5. Place the vials on ice for 2 min.

6. Add 250 μl of SOC medium to each vial.

7. Shake the vials horizontally at 37°C for 1 h at 225 r.p.m. in an rotary shaking incubator. Place the vials on ice.

8. Spread 50 μl and 100 μl from each transformation on separate LB agar medium containing 100 μl of ampicillin and X-gal.[a]

9. Invert the plates and incubate the plates at 37°C incubator for at least 18 h. Transfer the plates to 4°C for 2–3 h for blue colour development.

10. Pick up white colonies and screen recombinant clones by commercially available mini-prep plasmid isolation system (Qiagen or Jetprep). Dissolve the purified plasmid DNA in sterile H_2O instead of TE buffer.[b]

[a] Spread 20 μl of 50 mg/ml X-gal on each LB agar plate and incubate at 37°C for 30 min.

[b] The TE buffer (10 mM Tris–HCl pH 8.0, 1 mM EDTA) contains EDTA which may chelate Mg^{2+} needed by *Taq* DNA polymerase in the DNA sequencing reactions.

5.2 DNA sequencing

Use 0.5 μg of plasmid DNA containing the correct inserts for sequencing. The M13 reverse primer and the M13 forward primer can be used for sequencing the inserts from both directions. Sequencing can be performed using the AMI Prism dye terminator kit and an ABI 373 sequencer (Applied Biosystems).

5.3 Database searching

The nucleotide sequences obtained are compared with known sequences by searching GenBank and European Molecular Biology Laboratory databases with the *Fasta* Program (Genetic Computer Group).

A

```
A5   AGCCAGCGAA CTATACAACC TCCTAGGAAA GCTGTAGAGG AACCCCTTAA   50
A6   AGCCAGCGAA CTATACAACC TCCTAGGAAA GCTGTAGAGG AGCCCCTTAA
A7   AGCCAGCGAA CTATACAACC TCCTAGG--- ---------- ----------

A5   CGCGTTTAAA GNGTCGAAAG GGNTGATGAA TGACGAATAA CTGAAGGGAA   100
A6   CG-------- ---------- ---------- -----AATAA CTGAAGGGAA
A7   ---------- ---------- ---------- -----AATAA CTGAAGGGAA

A5   GTGACGGACT GTAACTTGAA AGTCAGAAGT GGAAGAATTC AGTTGTCTGA   150
A6   GTGACGGACT GTAACTTGAA AGTCAGAAGT GGAAGAATTC AGTTGTCTGA
A7   GTGACGGACT GTAACTTGAA AGTCAGAAGT GGAAGAATTC AGTTGTCTGA

A5   GCACCATGGC CTTCACAACT CACAGCTGGA AGGAACACAC CCCAGTGACT   200
A6   GCACCATGGC CTTCACAACT CACAGCTGGA AGGAACACAC AACAGTGACT
A7   GCACCATGGC CTTCACAACT CACAGCTGGA AGGAACACAC AACAGTGACT

A5   GACTTTCATC TCTGAAAAAA GTCATTAGAC CATAAATGAA TATTACAGTT   250
A6   GACTTTCATC TCTGAAAAAA GTCATTGAAC CATAAATGAA TATTACAGTT
A7   GACTTTCATC TCTGAAAAAA GTCATTGAAC CATAAATGAA TATTACAGTT

A5   AAGTTTATAT TAAAGCAGCT GTAATTTACG TAATAAAAAA TATATGATGT   300
A6   AAGTTTATAT TAAAGCAGCT GTAATTTACG TAATAAAAAA TATATGATGT
A7   AAGTTTATAT TAAAGCAGCT GTAATTTACG TAATAAAAAA TATATGATGT

A5   GCTGTGAAAA AAAAAAAA   318
A6   GCTGTGAAAA AAAAAAAA   285
A7   GCTGTGAAAA AAAAAAAA   260
```

B

```
upper     1 ACCAGGAATCTTGGCTCTGAATCCTTGGAATCAAGGAAATGACCTGTTCTCTCAAAGACAC
lower     1 ACCAGGAATCTTGGCTCTGAATCCTTGGAATCAAGGAAATGACCTGTTCTCTCAAAGACAC

upper    62 TGAAACAGGAGAGAGGGACTTCCATCCACTGGGCAGGGTACAGGCGCGTCTCAGTTGTGAA
lower    62 TGAAACAGGAGAGAGGGACTTCCATCCACTGGGCAGGGTACAGGCGCGTCTCAGTTGTGAA
                                                              **
upper   123 GGTCTATTCCTGGTTGCTCAGTCCCCAACct CGCATCACCCTGGGaccgccgCTTCTCAAC
lower   123 GGTCTATTCCTGGTTGCTCAGTCCCCAACTGCGCATCACCCTGGG         CTTCTCAAC
                 *
upper   184 CTGaAAGAGTCCACAACCATCCTTCTGAGGCCCTCCATCCCCACAACCACTAGCTGTTGTT
lower   177 CTGGAAGAGTCCACAACCATCCTTCTGAGGCCCTCCATCCCCACAACCACTAGCTGTTGTT

upper   245 CTCCAAGCCAAGGGCCCATTCCCTTTCTTATGCATGTACGGAGTATCGCCTAGACTTTAAG
lower   238 CTCCAAGCCAAGGGCCCATTCCCTTTCTTATGCATGTACGGAGTATCGCCTAGACTTTAAG
                                                            *
upper   306 CGTCCATCCTGTTTGAAAGTTTGGGAAACTGACACACGTTGTGTTCAAcCAGCCTGGTGTG
lower   299 CGTCCATCCTGTTTGAAAGTTTGGGAAACTGACACACGTTGTGTTCAAGCAGCCTGGTGTG

upper   367 GAGTGCCTTCGTATTAGTGTACCCTCTCGGAAGCTGGTTGGTGGGCAGGTGAGGAAGAAAT
lower   360 GAGTGCCTTCGTATTAGTGTACCCTCTCGGAAGCTGGTTGGTGGGCAGGTGAGGAAGAAAT
                                                           *      *
upper   428 GGAGCTGAAAGTGTCCCCTCAGTTGTCCTTTCCTCCCCCTCTAAGGTCCCTCcCTTTcCCC
lower   421 GGAGCTGAAAGTGTCCCCTCAGTTGTCCTTTCCTCCCCCTCTAAGGTCCCTCaCTTTtCCC

upper   489 AGGACATCGTACACTCCCCCCCCTTGTCACCTCTGCTAACATCTCAGAGCCAAGATTCCTGG
lower   482 AGGACATCGTACACTCCCCCCCCTTGTCACCTCTGCTAACATCTCAGAGCCAAGATTCCTGG

upper   550 T
lower   543 T
```

Figure 2 Sequence analysis of the differentially expressed cDNA. (A) Sequence of the A5, A6, and A7 cDNA fragments in *Figure 1A*. A5 is 33 bases longer than A6, and A6 is 25 bases longer than A7 (Huang and Lee, unpublished results). (B) Sequence of the upper and lower cDNA bands in *Figure 1B*. A seven base insertion is found in the upper cDNA band. (From ref. 8.)

Sequence comparison reveals that these three cDNA bands in *Figure 1A* are identical in their 3′ end cDNA sequences. However, at the 5′ end, A5 is 33 bp longer than A6, and A6 is 25 bp longer than A7. Database search results indicate the sequences are 80% and 70% homologous to the 3′ end region of the mouse integrin-associated protein (IAP) cDNA (accession number Z25524) and human IAP cDNA (accession number Z25521), respectively. It appears that A5, A6, and A7 cDNAs correspond to different alternative splicing forms of rat IAP mRNA. Sequence analysis from the cloned upper and lower bands in *Figure 1B* shows that these two bands share 98% homology in their cDNA sequences (*Figure 2B*). The sequences are found to be highly homologous (94–96% homology) with respect to the sequence of the 3′ untranslated region of the rat glial fibrillary acid protein (GFAP) cDNA. We have determined that these two cDNAs correspond to different isotypes of the GFAP mRNA and are results from the polymorphism of the rat GFAP gene (8).

6 Full-length cDNA cloning

The full-length cDNA can be cloned by screening a cDNA library (14) or by the rapid amplification of cDNA ends (RACE) method (15). We use the Marathon™ cDNA Amplification Kit (Clontech) to perform the RACE cloning. Procedures are performed according to the manufacturer's instructions. *Figure 3* shows the sequence of the rat IAP full-length cDNA.

7 Initial confirmation of PCR differentially expressed genes

7.1 Northern blot analysis

Northern blot analysis is recommended as the gold standard method to confirm the result of the PCR differential display analysis (10–13). The probes for Northern blot analysis can be obtained from three sources:

(a) Reamplified cDNA.

(b) Cloned cDNA obtained by reamplification.

(c) Sequence derived from the comparison of the differentially expressed cDNAs.

Using reamplified cDNA or cloned cDNA obtained by reamplification for Northern blot is described elsewhere (10–13) and is not discussed here. For our GFAP study, oligonucleotide sequences were designed according to the sequences of the differentially expressed cDNAs shown in *Figure 2B*. Two oligonucleotides, U1 and L1, were labelled with dig-dUTP and used as probes for Northern blot analysis (*Protocol 10*). *Figure 4* shows the results of Northern blot analysis of isotypes of the GFAP mRNA among untrained, poor memory, and good memory rats.

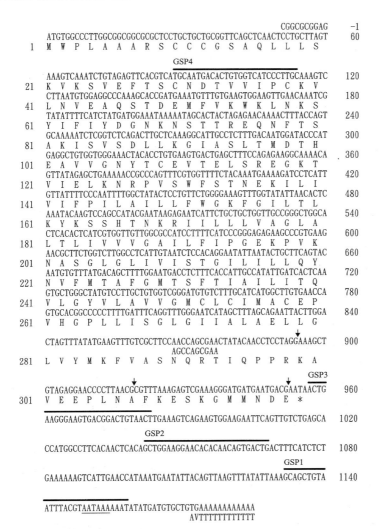

Figure 3 Nucleotide and deduced amino acid sequence of rat IAP cDNA. (From Ref. 6.)

Protocol 10

Northern blot analysis with non-isotope probes

Reagents

- 20 × SSC: 3 M NaCl, 0.3 M sodium citrate pH 7.0

- Maleic buffer: 0.1 M maleic acid, 0.15 M NaCl; adjust to pH 7.5 with solid NaOH

- Blocking solution: 1% (w/v) blocking reagent in maleic buffer

- Detection buffer: 0.1 M Tris–HCl, 0.1 M NaCl, 50 mM $MgCl_2$ pH 9.5

- DIG Oligonucleotide 3′ End Labelling Kit (Boehringer Mannheim)

- DIG DNA Labelling and Detection Kit (Boehringer Mannheim)

Protocol 10 continued

A. RNA gel electrophoresis and Northern transfer

1. Electrophorese 10 μg total hippocampal RNA through a 1% formaldehyde agarose gel containing 0.5 μg/ml ethidium bromide (16).

2. Photograph gel together with a ruler to allow subsequent assignment of mRNA size.

3. Rinse gel in 20 × SSC buffer at room temperature for 10 min.

4. Blot onto a nylon membrane (MSI) overnight and mark the wells by a pencil.

5. Cross-link the mRNA to the membrane by UV (5 J/cm^2).

B. Probe labelling

1. Use the DIG Oligonucleotide 3' End Labelling Kit (Boehringer Mannheim) to prepare DIG-dUTP labelled oligonucleotide probes following the manufacturer's instructions.

C. Pre-hybridization and hybridization

1. Pre-hybridize at 68°C for 2 h in 10 ml pre-hybridization solution containing 5 × SSC, 1% blocking solution, 0.1% N-lauryl sarcosine, 0.2% SDS.

2. Heat denature the probe by boiling for 10 min then cool on ice.

3. Hybridize at 68°C for 6–8 h in 8 ml pre-hybridization solution containing 10 pmol/ml of the denatured oligonucleotide probe.

D. Post-hybridization washing

1. Wash the filter twice with 2 × SSC, 0.1% SDS at 54°C for 15 min per wash.

2. Wash the filter twice with 0.1 × SSC, 0.1% SDS at 54°C for 15 min per wash.

E. Signal detection

Use the detection system such as DIG DNA Labelling and Detection Kit (Boehringer Mannheim) according the manufacturer's instructions.

1. Rinse the membrane briefly (1–5 min) in maleic acid buffer.

2. Incubate the membrane with blocking solution (100 ml/100 cm^2) at room temperature for 30 min.

3. Dilute anti-DIG–AP conjugate (1:20 000) in blocking solution.

4. Incubate membrane for 30 min in antibody solution (20 ml/100 cm^2) at room temperature for 30 min.

5. Wash 2 × 15 min with 100 ml maleic acid buffer.

6. Equilibrate 2–5 min in 20 ml detection buffer.

7. Dilute chemiluminescent reagent CDP-star™ in detection buffer (1:100).

8. Incubate membrane for 5 min in diluted CDP-star solution in a plastic bag.

9. Cut the plastic bag, dip on a Whatman membrane, and place the membrane in a clean plastic bag.

10. Expose to a ECL-Hyperfilm™ (Amersham) for 2–5 min.

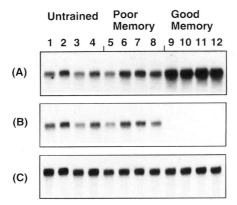

Figure 4 Northern blot analysis of isotypes of the GFAP mRNA. Four individuals each from the untrained (lanes 1–4), poor memory (lanes 5–8), and good memory (lanes 9–12) groups were analysed using 10 μg of total hippocampal RNA from each sample. The blot was hybridized with the L1 oligonucleotide probe first (B) followed by the U1 oligonucleotide probe (A), and finally hybridized with the rat GAPDH cDNA probe (C). (From ref. 8.)

7.2 Reverse Northern blot analysis

One of the drawbacks of Northern blot analysis is that large amount of RNA is needed. This is not feasible when the quantity of the RNA samples is limited. Recently, the reverse Northern blot analysis is used as an alternative approach to screen for cDNA fragments that truly represent differentially expressed mRNAs (17–19). Protocols for this method are presented elsewhere (18) and will not be described here.

7.3 Quantitative RT-PCR analysis

Another problem of Northern blot analysis is that of no signal on the blot when some of the differentially expressed cDNAs are used as probes. Since the PCR reactions in the differential display analysis run for 40 cycles, it is possible that the abundance of mRNA is too low to be detected by Northern blot analysis. For the IAP study, we perform semi-quantitative RT-PCR analyses (*Protocol 11*) for the confirmation of the differential gene expression.

Protocol 11

Quantitative RT-PCR analysis

Reagents

- See *Protocols 4–6*
- HPRT 1: 5'-CTC TGT GTG CTG AAG GGG GG-3'
- HPRT 2: 5'-GGG ACG CAG CAA CAG ACA TT-3'
- IAP 1: 5'-ATT TTT TAT TAC GTA AAT TAC AGC TGC-3'
- IAP 3: 5'-AAC TGA AGG GAA GTG ACG GAC TGT AAC-3'

Protocol 11 continued

A. Reverse transcription

1. To establish a standard curve, make a twofold dilution of a single RNA preparation (i.e. 0.125, 0.25, 0.5, 1.0, and 2.0 μg RNA). Add 1 μl of 0.5 μg/μl of oligo(dT) into each RNA sample. Add DEPC H$_2$O to make a final volume of 9 μl.

2. Incubate at 70°C for 10 min, set tubes on ice for 5 min.

3. Spin down, add 11 μl of a reverse transcription mixture containing:
   - 4 μl of 5 $\times$ reverse transcription buffer
   - 2 μl of 0.1 M DTT
   - 1 μl of 10 mM dNTP
   - 0.28 μl RNasin (40 U/μl)
   - 0.5 μl Superscript II MMLV (200 U/μl)
   - 3.22 μl DEPC H$_2$O

4. Incubate at 37°C for 10 min, terminate the reaction by heating at 80°C for 10 min. Add 30 μl of DEPC H$_2$O to the RT products for dilution.

B. PCR

1. Add 1/20 (2.5 μl) of the RT products to 15 μl of PCR reaction mix containing:
   - 1 $\times$ PCR buffer
   - 1.0 mM MgCl$_2$
   - 0.2 mM dNTP[a]
   - 1.0 μl [α-^{35}S]dATP
   - 1 U *Taq* DNA polymerase

2. Add 2.5 μl of the primer mix containing:
   - 0.1 μM HPRT 1
   - 0.1 μM HPRT 2
   - 0.2 μM IAP 1
   - 0.2 μM IAP 3

3. PCR parameters: 39 cycles of 94°C for 30 sec, 60°C for 1 min, and 72°C for 30 sec. Then a final elongation at 72°C for 5 min.

C. Polyacrylamide gel analysis

1. Prepare a native 9% polyacrylamide gel in 1 $\times$ TBE buffer.

2. Mix 18 μl of the PCR products with 2 μl of the 10 $\times$ loading dye.

3. Load the samples on the polyacrylamide gel.

4. Run the gel at 200 V constant voltage for 3–4 h.

5. Dry the gel and autoradiograph on an imaging plate of a PhospholImage analyser (PhospholImager, Molecular Dynamics).

6. Quantify the radioactivity of each cDNA band according to the instruction manual.

[a] When using [α-^{35}S]dATP for labelling PCR products, decrease the concentration of unlabelled dATP in the dNTP mix to 0.1 mM.

(A)

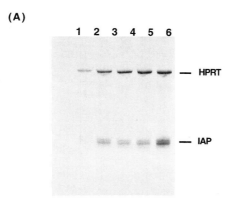

(B)

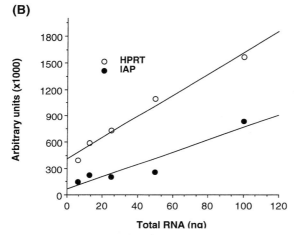

Figure 5 Quantitative RT-PCR analysis of rat IAP mRNA. (A) Autoradiograph of the IAP and HPRT cDNA bands. Serial quantities (6.25, 12.5, 25, 50, and 100 ng) of total hippocampal RNA were reverse transcribed and amplified by PCR. The template of HPRT was used as an

Figure 5 shows the results of the semi-quantitative RT-PCR. A linear relationship was established between serial amounts of hippocampal total RNA and the optical densities of the cDNA bands (*Figures 5A* and *5B*). Statistical analyses revealed that the IAP mRNA level was significantly higher in good memory rats when compared with the poor memory rats (t = 8.11; p < 0.01).

8 Further confirmation of PCR differentially expressed genes

The approaches we adopt to further confirm the differentially expressed gene associated with memory formation include:

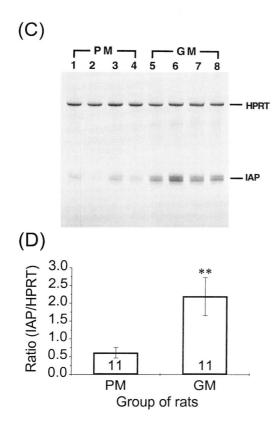

internal control. RT-PCR products were analysed on a 9% polyacrylamide gel, visualized by a PhosphoImager machine, and quantitated. (B) Linear relationship between the optical density of the cDNA bands and the quantity of total RNA. (C) Autoradiography of the IAP and HPRT cDNA bands from the poor memory (PM) and good memory (GM) rats. (D) Higher expression of IAP mRNA level in good memory rats (n = 11 in each group. ** $p < 0.01$ by Student's test). (From ref. 6.)

(a) *In situ* hybridization.

(b) Down-regulation of IAP gene expression by antisense oligonucleotide treatment.

(c) Using of IAP gene knock-out mice to study the relationship of IAP gene and memory formation.

8.1 *In situ* hybridization

In situ hybridization is used to detect the regional distribution of IAP mRNA in the rat brain (*Protocol 12*) and to further confirm the differential expression of IAP mRNA in the hippocampus of poor memory and good memory rats.

Protocol 12

In situ hybridization

Reagents

- Pentobarbital
- 0.1 M phosphate buffer saline (PBS) pH 7.4
- 4% paraformaldehyde in 0.1 M PBS
- Poly-L-lysine coated slides
- 10 × tailing buffer
- Yeast transfer RNA

- Terminal deoxynucleotidyl transferase (25 U/μl)
- Salmon sperm DNA
- Denhardt's solution: 0.02% Ficoll, 0.02% polyvinyl pyrrolidone, 0.02% BSA

A. Retention test, animal perfusion, and sectioning

1. 3 h after training, rats showing poor memory or good memory are anaesthetized with pentobarbital (40 mg/kg, i.p.).

2. Rats are then perfused intracardially in a chemical hood with 150 ml of heparinized 0.1 M PBS, followed by 150 ml of 4% paraformaldehyde in 0.1 M PBS.

3. Remove the brain and store at −80°C until sectioning.

4. Serial sections at 20 μm thickness through the hippocampus are cut on a cryostat, thaw-mounted onto poly-L-lysine coated slides, and vacuum desiccated overnight.

5. Store the slides in boxes containing desiccant at −80°C until *in situ* hybridization is performed.

B. Oligonucleotide probe labelling

1. A 46 base synthetic oligonucleotide (5′-CCA CTT CAC AAA CAT TTC ATC GGT GCT TTG GGC CTC CAC ATT AAG G-3′) complementary to the cloned rat IAP cDNA (bases 129–174) was used. The oligonucleotide was synthesized and purified by Genosys Biotechnologies Inc.

2. In a sterile tube, add the following:
   - 1 μl oligonucleotide (15 pmole)
   - 1.25 μl of 10 × tailing buffer
   - 8.25 μl DEPC H₂O
   - 1 μl terminal deoxynucleotidyl transferase (25 U/μl)

3. Mix gently by pipetting. Incubate at 37°C for 15 min.

4. Unincorporated nucleotides are separated from labelled probe by a spin column procedure using Sephadex G25. The final specific activity of the probe is about 10^6 c.p.m./μl.

C. Hybridization, post-hybridization washing, and development

1. For hybridization, apply the labelled IAP oligonucleotide probe (1×10^6 c.p.m./slide) in 100 μg/ml yeast transfer RNA, 500 μg/ml salmon sperm DNA, and Denhardt's solution to each slide.

Protocol 12 continued

2. Cover the slides with Parafilm, and hybridize for 24 h at 42 °C.

3. Remove the coverslips and rinse the sections with 2 × SSC, then with 1 × SSC containing 1 M DTT (0.1%), followed by 30 min wash in 0.5 × SSC containing 1 M DTT at 47 °C. A final wash in 0.5 × SSC containing 1 M DTT is performed at room temperature for 30 min.

4. Dehydrate the slides through a series of ethanol and dip in emulsion (NTB-3, Kodak) diluted 1:1 with distilled water.

5. After a three week exposure period, develop the slides in Kodak D-19 developer.

Figure 6 shows the results of the *in situ* hybridization. IAP mRNA is expressed in both the pyramidal cell layer and the dentate gyrus of the hippocampus. The IAP mRNA level is higher in both subdivisions of the hippocampus in good memory rats (*Figure 6B*) when compared with the poor memory controls (*Figure 6A*), but the effect is more prominent in the dentate gyrus (7).

(A) Poor memory

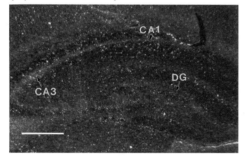

(B) Good memory

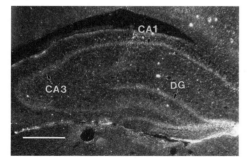

Figure 6 *In situ* hybridization showing a higher expression of IAP mRNA signal in the hippocampus of good memory rats. Coronal sections through the hippocampus from (A) poor memory rats (n = 4) and (B) good memory rats (n = 6) were subjected to *in situ* hybridization analyses. CA1, CA1 cell body layer; CA3, CA3 cell body layer; DG, dentate gyrus. Scale bar, 500 μm. (From ref. 6.)

8.2 Gene expression down-regulated by antisense oligonucleotide treatment

If IAP is related to memory formation, blocking the expression of IAP gene should impair memory and memory-related functions. We carried out an antisense oligonucleotide study to test this hypothesis. Antisense oligonucleotides were designed, synthesized, and injected into the rat hippocampus (*Protocol 13*). The effects of the antisense oligonucleotide are examined in the behavioural task as well as in the long-term potentiation (LTP) paradigm (*Protocol 14*).

Protocol 13

Antisense oligonucleotide study

A. Design and synthesis of the antisense oligonucleotide

1. The antisense oligonucleotide for IAP is an 18-mer, and the sequence is 5′-CGC CGC CAA GGG CCA CAT-3′, which is complementary to nucleotides 1–18 of the sequence of rat IAP cDNA.

2. The antisense oligonucleotide is a phosphothioate derivative.

3. A random 18-mer sequence 5′-TGA GAA GAG TGA TGA CAA-3′ was synthesized as a control.

B. Antisense oligonucleotide administration

1. For the memory retention experiment, rats were randomly divided into three groups (n = 11 or 9) (see *Figure 7*). Rats in all groups were subjected to stereotaxic surgery and intra-dentate injections as described above. For this experiment, 1 nmole of IAP antisense or random sequence oligonucleotide in 1.0 μl was injected to each rat for four times, with 12 h apart between injections, at a rate of 0.2 μl/min. Rats were trained 12 h after the last injection and memory retention measured 3 h after training.

2. For the LTP experiment (*Protocol 14*), the IAP antisense or the random sequence was similarly injected into one side of the dentate area at a position of 1.0 mm above the gyrus. Two injections (1 nmole) were given with the first injection 16 h and the second injection 2 h before the electrophysiological recording. For each injection, 1.0 μl of 1.0 nmole antisense or random sequence was administered at a rate of 0.2 μl/min.

As shown in *Figure 7*, IAP antisense oligonucleotide administration significantly impaired retention performance in rats (Mann-Whitney U test, U = 18.5, Z = 2.35, p < 0.05 when compared with the controls). Injection of random sequence oligonucleotide did not produce a marked effect on memory retention (Mann-Whitney U test, U = 58.5, Z = 0.13, p > 0.05 when compared with the controls).

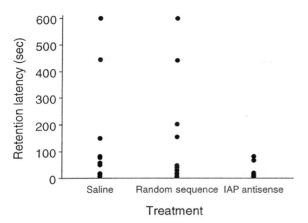

Figure 7 Effects of IAP antisense oligonucleotide on memory retention of one-way inhibitory avoidance learning in rats. The distribution of the retention score for each individual rat upon saline (n = 11), IAP antisense oligonucleotide (n = 11), or random sequence (n = 9) treatment is shown. Four injections at intervals of 12 h were given before the training procedure (see *Protocol 13* for details). For each injection, 1.0 μl of saline or the oligonucleotide (1 nmole) was directly injected into the dentate gyrus of the hippocampus bilaterally. There was a significant difference between the control group and the IAP antisense group when evaluated by the Mann-Whitney U test (p < 0.05). (From ref. 6.)

Protocol 14

In vivo LTP study

Equipment and reagents

- Sprague-Dawley rats with body weight 250–350 g
- Stereotaxic instrument
- Stimulating electrodes: platinum concentric bipolar electrodes
- Surgery instrument
- Recording electrodes: glass electrodes filled with 3 M NaCl
- Extracellular recording system
- 5% agar dissolved in 0.9% NaCl

Method

1. Inject urethane (1.4 g/kg) intraperitoneally to anaesthetize rat and place the rat on a stereotaxic instrument.

2. Throughout the surgery and experiment, core body temperature is monitored and maintained at 35 ± 1 °C with a feedback control system.

3. Expose the skull and identify the bregma.

4. Prepare the recording electrodes by pulling a single barrel glass micropipette (1.2 mm o.d. × 0.6 mm i.d.) on a Narishige vertical puller and filled with 3 M NaCl. Resistance ranges from 1–3 MΩ.

5. Implant the recording electrodes ipsilaterally into the dentate gyrus, 3.5 mm posterior to the bregma and 2.0 mm lateral to the midline.

Protocol 14 continued

6. Implant the stimulating electrodes unilaterally to the dorsomedial perforant path at stereotaxic coordinates of 8.5 mm posterior and 4.4 mm lateral to the bregma. The stereotaxic coordinates are adjusted for variation in rat body weights and to maximize the monosynaptic responses of the population excitatory postsynaptic potentials (pEPSP) produced by the granular cells in response to stimulation of the perforant path.

7. Once both the recording and stimulating electrodes are positioned, apply 2–3 ml of melted 5% agar over the exposed skull to prevent surface drying and reduce movement artefacts.

8. Stimulation consists of 50 μsec duration monophasic constant current pulses delivered once per 10 min. Stimulus intensities range from 50 μA to 250 μA and produce averaged pEPSP amplitudes of 3–5 mV. Once determined, stimulus current remains constant throughout the experiment.

9. To induce LTP, four sets of stimulus trains in a 10 min period are delivered following a 30 min baseline recording. Each set contains five trains, ten pulses per train at 400 Hz, delivered at a rate of one train per sec for 5 sec. The pulse widths in the trains are 50, 100, 150, and 200 μsec, respectively. The population spike amplitude, slope, and amplitude of the population excitatory postsynaptic potential (pEPSP) are recorded once per 10 min.

As shown in *Figure 8*, IAP antisense oligonucleotide injection to the dentate gyrus markedly decreased the expression of *in vivo* LTP in both the amplitude (41% decrease on the average, $tD = 3.61$, $p < 0.01$, Dunnett's t-test) (*Figure 8A*) and slope (37% decrease on the average, $tD = 3.28$, $p < 0.01$) (*Figure 8B*) of pEPSP in the hippocampus when compared with the tetanization group. There was not a marked difference between the tetanization group and the random sequence control group as the amplitude of pEPSP ($tD = 0.91$, $p > 0.05$) and slope of pEPSP ($tD = 0.52$, $p > 0.05$) are concerned.

8.3 Gene knock-out study

Other than the antisense oligonucleotide treatment and the LTP paradigm mentioned above, a direct method to confirm the observed effects of IAP involved in memory formation is to adopt the gene knock-out (KO) approach. We have therefore used the IAP-deficient mice (20) to examine their memory retention performance. Our results show that there is a significant impairment in retention performance in IAP-KO mice ($IAP^-/^-$, $n = 17$) when compared with either the homozygote $IAP^+/^+$, $n = 13$) or heterozygote ($IAP^\pm$, $n = 11$) controls (Mann-Whitney U test, $Z = 1.74$ and $Z = 2.14$, respectively, both $p < 0.05$). While there was not a marked difference in retention performance between the homozygote and heterozygote animals (Mann-Whitney U test, $Z = 0.38$, $p > 0.05$) (*Figure 9A*). One would suspect that retention performance may be affected by factors other than memory consolidation, such as motor activity and

(A)

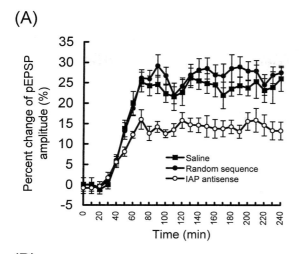

(B)

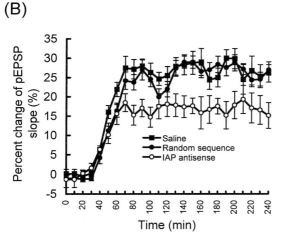

Figure 8 Effects of IAP antisense oligonucleotide on long-term potentiation in rat hippocampus. Per cent change in the (A) mean amplitude and (B) slope of pEPSPs and SEMs was presented as a function of time upon IAP antisense oligonucleotide injection (n = 5) and the random sequence injection (n = 5) versus the tetanization controls (n = 5). The tetanus stimulations which yielded LTP (indicated by *arrows*) were given at 30, 40, 50, and 60 min. Each stimulation contains five trains at 400 Hz. The duration of stimulation was 50 μsec at 30 min, 100 μsec at 40 min, 150 μsec at 50 min, and 200 μsec at 60 min, respectively. (From ref. 6.)

balance. Further analyses revealed that the IAP-KO mice were no different from the homozygote and heterozygote controls when compared separately in locomotor activity (Dunnett's t-test followed one-way analysis of variance (ANOVA), tD = 0.97, p > 0.05, and tD = 0.64, p > 0.05, respectively) (*Figure 9B*) and running wheel activity (tD = 0.10, p > 0.05, and tD = 1.0, p > 0.05, respectively) assays performed after the retention test (*Figure 9C*). We have also examined whether these animals may be different in their sensitivities to the

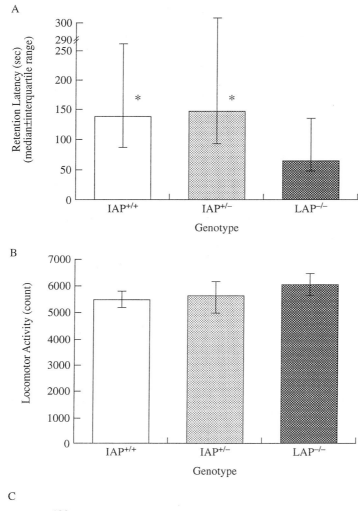

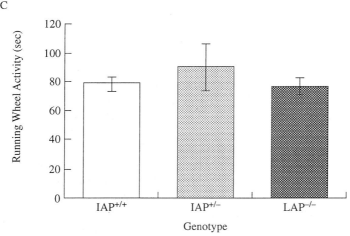

Figure 9 (A) Retention performance of homozygote ($IAP^{+/+}$, n = 13), heterozygote ($IAP^{+/-}$, n = 11), and IAP-KO ($IAP^{-/-}$, n = 17) mice in an inhibitory avoidance learning task. Data are expressed as median $\pm$ interquartile range. The comparisons between the $IAP^{+/+}$ and $IAP^{-/-}$ groups as well as between the $IAP^{+/-}$ and $IAP^{-/-}$ groups were both significant (both $p < 0.05$, Mann-Whitney one-tailed U-test). (B) Locomotor activity level of the same mice measured in a digiscan activity monitor for 20 min ($p > 0.05$). (C) Running wheel activity if the same animals measured in a rota-rod treadmill ($p > 0.05$). Data are expressed as mean $\pm$ SEM in (B) and (C). (From ref. 21.)

electric shock. We have therefore conducted an experiment to have all animals exposed to different intensities of the electric shock and their responses were recorded. *Table 2* illustrates that there was indeed no marked difference in both response category and percentage of animals showing a specific response to electric shocks among IAP-KO, homozygote, and heterozygote mice. These results demonstrate that the observed difference in retention performance is not due to different sensitivities of these animals to the electric shock.

Table 2 Responses of $IAP^{+/+}$, $IAP^{+/-}$, and $IAP^{-/-}$ mice to different shock intensities[a] (From ref. 21).

| Genotype[b] | n[c] | Shock intensity (mA) | | | | |
		0	0.4	0.8	1.2	1.6
$IAP^{+/+}$	13	–	F (17%), L (63%), V	L, J, V	J, V	J, V
$IAP^{+/-}$	11	–	F (14%), L (65%), V	L, J, V	J, V	J, V
$IAP^{-/-}$	17	–	F (18%), L (53%), V	L, J, V	J, V	J, V

[a] F: flinch, L: locomotion, V: vocalization, J: jump. The value in parenthesis indicates the percentage of animals showing that specific response category. Response category without percentage expression indicates all animals in that group show that behavioural response.

[b] There is no marked difference in any behavioural response observed among $IAP^{+/+}$, $IAP^{+/-}$, and $IAP^{-/-}$ mice at any shock intensity examined.

[c] n: number of animals.

9 Conclusions

By using the differential display method and by adopting the one-way inhibitory avoidance learning paradigm, we have identified at least two genes which are specifically associated with memory processing in rats. The first gene is integrin-associated protein (IAP), and its gene expression is significantly increased in rats showing good retention performance three hours post-training. This finding has been confirmed by *in situ* hybridization examination, antisense oligonucleotide treatment, and gene knock-out study. *In vivo* LTP recording, which reflects a form of synaptic plasticity, also supports the above findings. Moreover, the expression of the IAP gene belongs to neuronal plasticity in that IAP expression also exists in poor memory and control rats, but its expression is significantly enhanced during the memory consolidation process upon learning in some animals (animals showing good retention performance). The second gene is glial fibrillary acid protein (GFAP). Rats showing good retention performance 24 hours post-training preferentially express one isoform of the GFAP gene while rats showing poor retention performance and control rats preferentially express the other isoform of the same gene. Hence, in contrast to neuronal plasticity, the relationship between GFAP gene expression and memory performance belongs to DNA polymorphism. Learning and memory formation are complex behaviours involving complicated neuronal processes. Different populations of neurons and neuronal pathways are probably involved in different forms of learning and memory. Even in the same form of learning, different neuronal groups are also possibly involved at different stages of the memory process. Therefore, various gene expressions, the signal transduction pathways associated with these gene expressions, and the regulations of these gene expressions may also differ depending upon the learning paradigm and the stage of memory processing are involved.

Acknowledgements

Dr A.-M. Huang is an Assistant Professor at the Department of Physiology, National Cheng-Kung University, Tainan, Taiwan. Dr Eminy H. Y. Lee is a Research Fellow at the Institute of Biomedical Sciences, Academia Sinica, Taipei, Taiwan, Republic of China. This work was supported by Research Funds from the Institute of Biomedical Sciences, Academia Sinica, Taiwan and partly by a grant from the National Science Council of Taiwan (NSC88–2312-B001–013), Republic of China.

References

1. Davis, H. P. and Squire, L. R. (1984). *Psychol. Bull.*, **96**, 518.
2. Lee, E. H. Y., Hung, H. C., Lu, K. T., Chen, W. H., and Chen, H. Y. (1992). *Peptides*, **13**, 927.
3. Castellucci, V. F., Kennedy, T. E., Kandel, E. R., and Goelet, P. (1988). *Neuron*, **1**, 321.
4. Tully, T. (1996). *Proc. Natl. Acad. Sci. USA*, **93**, 13460.
5. Liang, P. and Pardee, A. B. (1992). *Science*, **257**, 967.
6. Huang, A. M., Wang, H. L., Tang, Y. P., and Lee, E. H. Y. (1998). *J. Neurosci.*, **18**, 4305.
7. Bliss, T. V. P. and Collingridge, G. L. (1993). *Nature*, **361**, 31.
8. Huang, A. M. and Lee, E. H. Y. (1997). *NeuroReport*, **8**, 1619.
9. Squire, L. R. and Alvarez, P. (1995). *Curr. Opin. Neurobiol.*, **5**, 169.
10. Liang, P., Averboukh, L., and Pardee, A. B. (1993). *Nucleic Acids Res.*, **21**, 3269.
11. Zhu, W. and Linag, P. (1996). In *Methods in molecular biology*, Vol. 68: *Gene isolation and mapping protocols* (ed. J. Boultwood). Humana Press Inc., Totowa, NJ.
12. Liang, P. and Pardee, A. B. (1997). In *Methods in molecular biology*, Vol. 85: *Differential display methods and protocols* (ed. P. Liang and A. B. Pardee). Human Press Inc., Totowa, NJ.
13. Liang, P. and Pardee, A. B. (1995). *Curr. Opin. Immunol.*, **7**, 274.
14. Ausubel, F., Brent, R., Kingston, R. E., Moore, D. D., Seidman, J. G., Smith, J. A., *et al.* (1988). *Current protocols in molecular biology*. Greene and Wiley-Interscience, New York.
15. Frohman, M. A., Dush, M. K., and Martin, G. R. (1988). *Proc. Natl. Acad. Sci. USA*, **85**, 8998.
16. Sambrook, J., Fritsch, E. F., and Maniatis, T. (ed.) (1989). *Molecular cloning: a laboratory manual*, 2nd edn. Cold Spring Harbor Laboratory Press, Cold Spring Harbor, NY.

17. Mou, L., Miller, H., Li, J., Wang, E., and Chalifour, L. (1994). *Biochem. Biophys. Res. Commun.*, **199**, 564.

18. Zhang, H., Zhang, R., and Liang, P. (1997). In *Methods in molecular biology*, Vol. 85: *Differential display methods and protocols* (ed. P. Liang and A. B. Pardee). Human Press Inc., Totowa, NJ.

19. Martin, K. J., Dwan, C., O'Hare, M. J., Pardee, A. B., and Sager, R. (1998). *BioTechniques*, **24**, 1018.

20. Lindberg, F. P., Bullard, D. C., Caver, T. E., Gresham, H. D., Beaudet, A. L., and Brown, E. J. (1996). *Science*, **274**, 795.

21. Chang, H. P., Lindberg, F. P., Wang, H. L., Huang, A. M. and Lee, E. H. Y. (1999). *Learn. Mem.*, **6**, 448.

List of suppliers

Ambion, Inc., 2130 Woodward Street, 200, Austin, TX 78744-1832, USA

Amresco, 30175 Solon Industrial Parkway, Solon, OH 44139, USA

Anderman and Co. Ltd, 145 London Road, Kingston-upon-Thames, Surrey, KT2 6NH

Tel: 0181 541 0035   Fax: 0181 541 0623

Applied Scientific, 154 W Harris Avenue, South San Francisco, CA 94080, USA

Beckman Coulter Inc., 4300 N Harbor Boulevard, PO Box 3100, Fullerton, CA 92834-3100, USA

Tel: 001 714 871 4848   Fax: 001 714 773 8283   Web site: www.beckman.com

Beckman Coulter (U.K.) Limited, Oakley Court, Kingsmead Business Park, London Road, High Wycombe, Buckinghamshire, HP11 1JU

Tel: 01494 441181   Fax: 01494 447558   Web site: www.beckman.com

Becton Dickinson and Co., 21 Between Towns Road, Cowley, Oxford, OX4 3LY

Tel: 01865 748844   Fax: 01865 781627   Web site: www.bd.com

Becton Dickinson and Co., 1 Becton Drive, Franklin Lakes, NJ 07417-1883, USA

Tel: 001 201 847 6800   Web site: www.bd.com

BioDiscovery, 11150 W Olympic Blvd. Ste. 805E, Los Angeles, CA 90064, USA

Bio 101 Inc., c/o Anachem Ltd, Anachem House, 20 Charles Street, Luton, Bedfordshire, LU2 0EB

Tel: 01582 456666   Fax: 01582 391768   Web site: www.anachem.co.uk

Bio 101 Inc., PO Box 2284, La Jolla, CA 92038-2284, USA

Tel: 001 760 598 7299   Fax: 001 760 598 0116   Web site: www.bio101.com

Bio-Rad Laboratories Ltd, Bio-Rad House, Maylands Avenue, Hemel Hempstead, Hertfordshire, HP2 7TD

Tel: 0181 328 2000   Fax: 0181 328 2550   Web site: www.bio-rad.com

Bio-Rad Laboratories Ltd., Division Headquarters, 1000 Alfred Noble Drive, Hercules, CA 94547, USA

Tel: 001 510 724 7000   Fax: 001 510 741 5817   Web site: www.bio-rad.com

Cartesian Technologies Inc., 17781 Sky Park Circle, Irvine, CA 92614, USA

CP Instrument Company Ltd, PO Box 22, Bishop Stortford, Hertfordshire, CM23 3DX

Tel: 01279 757711   Fax: 01279 755785   Web site: www.cpinstrument.co.uk

Dupont (UK) Ltd, Industrial Products Division, Wedgwood Way, Stevenage, Herts, SG1 4QN

Tel: 01438 734000   Fax: 01438 734382   Web site: www.dupont.com

Dupont Co, (Biotechnology Systems Division), PO Box 80024, Wilmington, DE 19880-002, USA

Tel: 001 302 774 1000   Fax: 001 302 774 7321   Web site: www.dupont.com

Eastman Chemical Company, 100 North Eastman Road, PO Box 511, Kingsport, TN 37662-5075, USA

Tel: 001 423 229 2000   Web site: www.eastman.com

Fisher Scientific UK Ltd, Bishop Meadow Road, Loughborough, Leicestershire, LE11 5RG

Tel: 01509 231166   Fax: 01509 231893   Web site: www.fisher.co.uk

Fisher Scientific, Fisher Research, 2761 Walnut Avenue, Tustin, CA 92780, USA

Tel: 001 714 669 4600   Fax: 001 714 669 1613   Web site: www.fishersci.com

Fluka, P.O. Box 2060, Milwaukee, WI 53201, USA

Tel: 001 414 273 5013   Fax: 001 414 2734979   Web site: www.sigma-aldrich.com

Fluka Chemical Company Ltd, PO Box 260, CH-9471, Buchs, Switzerland

Tel: 00 41 81 745 2828   Fax: 00 41 81 756 5449   Web site: www.sigma-aldrich.com

GeneMachines, PO Box 2048, Menlo Park, CA 94026, USA

General Scanning Inc., 500 Arsenal Street, Watertown, MA 02472, USA

Genetic Microsystems Inc., 34 Commerce Way, Woburn, MA 01801, USA

Genetix Ltd, Unit 1, 9 Airfield Road, Christchurch, Dorset BH23 3TG

GenHunter Corp., 624 Grassmere Park Drive, Suite 17, Nashville, TN 37211, USA

Tel: 615 833 0665   Fax: 615 832 9461

Genomic Solutions Inc., 4355 Varsity Drive Ste. E, Ann Arbor, MI 48108, USA

Genome Systems Inc., 4633 World Parkway Circle, St. Louis, MO 63134, USA

Genosys Biotechnologies Inc., Lake Front Circle, Suite 185, The Woodlands, TX 77380, USA

Hybaid Ltd, Action Court, Ashford Road, Ashford, Middlesex, TW15 1XB

Tel: 01784 425000   Fax: 01784 248085   Web site: www.hybaid.com

Hybaid US, 8 East Forge Parkway, Franklin, MA 02038, USA

Tel: 001 508 541 6918   Fax: 001 508 541 3041   Web site: www.hybaid.com

HyClone Laboratories, 1725 South HyClone Road, Logan, UT 84321, USA

Tel: 001 435 753 4584   Fax: 001 435 753 4589   Web site: www.hyclone.com

Imaging Research Inc., Brock Univesity, 500 Glenridge Avenue, St. Catherines, Ontario L2S 3A1, Canada

Intelligent Automation Systems, 149 Sidney Street, Cambridge, MA 02139, USA

IVEE Development AB, Forsta Langgata 26, SE-413 28 Goteborg, Sweden

Invitrogen BV, PO Box 2312, 9704 CH Groningen, The Netherlands

Tel: 00800 5345 5345   Fax: 00800 7890 7890   Web site: www.invitrogen.com

Invitrogen Corporation, 1600 Faraday Avenue, Carlsbad, CA 92008, USA

Tel: 001 760 603 7200   Fax: 001 760 603 7201   Web site: www.invitrogen.com

Life Technologies Ltd, PO Box 35, Free Fountain Drive, Incsinnan Business Park, Paisley, PA4 9RF

Tel: 0800 269210   Fax: 0800 838380   Web site: www.lifetech.com

Life Technologies Inc, 9800 Medical Center Drive, Rockville, MD 20850, USA
Tel: 001 301 610 8000   Web site: www.lifetech.com
Merck Sharp & Dohme Research Laboratories, Neuroscience Research Centre, Terlings Park, Harlow, Essex CM20 2QR
                                         Web site: www.msd-nrc.co.uk
Microflex, PO Box 1865, San Francisco, CA 94083-1865, USA
MSD Sharp and Dohme GmbH, Lindenplatz 1, D-85540, Haar, Germany
                                         Web site: www.msd-deutschland.com
Millipore (UK) Ltd, The Boulevard, Blackmoor Lane, Watford, Hertfordshire, WD1 8YW
Tel: 01923 816375   Fax: 01923 818297
                                 Web site: www.millipore.com/local/UK.htm
Millipore Corporation, 80 Ashby Road, Bedford, MA 01730, USA
Tel: 001 800 645 5476   Fax: 001 800 645 5439   Web site: www.millipore.com
Molecular Dynamics, 928 East Arques Avenue, Sunnyvale, CA 94086, USA
Molecular Probes, Inc., 4849 Pitchford Avenue, PO Box 22010, Eugene OR 97402-9165, USA
New England Biolabs, 32 Tozer Road, Beverley, MA 01915-5510, USA
Tel: 001 978 927 5054
Nikon Corporation, Fuji Building, 2-3, 3-chome, Marunouchi, Chiyoda-ku, Tokyo 100, Japan
Tel: 00 813 3214 5311   Fax: 00 813 3201 5856
                         Web site: www.nikon.co.jp/main/index_e.htm
Nikon Inc, 1300 Walt Whitman Road, Melville, NY 11747-3064, USA
Tel: 001 516 547 4200   Fax: 001 516 547 0299   Web site: www.nikonusa.com
Nycomed Amersham plc, Amersham Place, Little Chalfont, Buckinghamshire, HP7 9NA
Tel: 01494 544000   Fax: 01494 542266   Web site: www.amersham.co.uk
Nycomed Amersham, 101 Carnegie Center, Princeton, NJ 08540, USA
Tel: 001 609 514 6000   Web site: www.amersham.co.uk
Perkin Elmer Ltd, Post Office Lane, Beaconsfield, Buckinghamshire, HP9 1QA
Tel: 01494 676161   Web site: www.perkin-elmer.com
Pharmacia Biotech (Biochrom) Ltd, Unit 22, Cambridge Science Park, Milton Rd, Cambridge, Cambs, CB4 0FJ
Tel: 01223 423723   Fax: 01223 420164   Web site: www.biochrom.co.uk
Pharmacia and Upjohn Ltd, Davy Avenue, Knowlhill, Milton Keynes, Buckinghamshire, MK5 8PH
Tel: 01908 661101   Fax: 01908 690091   Web site: www.eu.pnu.com
Promega Corporation, 2800 Woods Hollow Road, Madison, WI 53711-5399, USA
Tel: 001 608 274 4330   Fax: 001 608 277 2516   Web site: www.promega.com
Promega UK Ltd, Delta House, Chilworth Research Centre, Southampton, SO16 7NS
Tel: 0800 378994   Fax: 0800 181037   Web site: www.promega.com
Qiagen UK Ltd, Boundary Court, Gatwick Road, Crawley, West Sussex, RH10 2AX
Tel: 01293 422911   Fax: 01293 422922   Web site: www.qiagen.com

Qiagen Inc, 28159 Avenue Stanford, Valencia, CA 91355, USA

Tel: 001 800 426 8157   Fax: 001 800 718 2056   Web site: www.qiagen.com

Research Genetics Inc., 2130 Memorial Pkwy SW, Huntsville, AL 35801, USA

Roche Diagnostics Ltd, Bell Lane, Lewes, East Sussex, BN7 1LG

Tel: 01273 484644   Fax: 01273 480266   Web site: www.roche.com

Roche Diagnostics Corporation, 9115 Hague Road, PO Box 50457, Indianapolis, IN 46256, USA

Tel: 001 317 845 2358   Fax: 001 317 576 2126   Web site: www.roche.com

Roche Diagnostics GmbH, Sandhoferstrasse 116, 68305 Mannheim, Germany

Tel: 0049 621 759 4747   Fax: 0049 621 759 4002   Web site: www.roche.com

Schleicher and Schuell Inc., Keene, NH 03431A, USA

Tel: 001 603 357 2398

Shandon Scientific Ltd, 93-96 Chadwick Road, Astmoor, Runcorn, Cheshire, WA7 1PR

Tel: 01928 566611   Web site: www.shandon.com

Sigma-Aldrich Company Ltd, Fancy Road, Poole, Dorset, BH12 4QH

Tel: 01202 722114   Fax: 01202 715460   Web site: www.sigma-aldrich.com

Sigma Chemical Company, PO Box 14508, St Louis, MO 63178, USA

Tel: 001 314 771 5765   Fax: 001 314 771 5757   Web site: www.sigma-aldrich.com

Stratagene Europe, Gebouw California, Hogehilweg 15, 1101 CB Amsterdam Zuidoost, The Netherlands

Tel: 00 800 9100 9100   Web site: www.stratagene.com

Stratagene Inc, 11011 North Torrey Pines Road, La Jolla, CA 92037, USA

Tel: 001 858 535 5400   Web site: www.stratagene.com

Synteni (Incyte Pharmaceuticals), 6519 Dumbarton Circle, Fremont, CA 94555, USA

Tel-Test, Inc., 1511 County Road 129 Po Box 1421, Friendswood, TX 77546, USA

United States Biochemical, PO Box 22400, Cleveland, OH 44122, USA

Tel: 001 216 464 9277

Vysis Inc., 3100 Woodcreek Drive, Downers Grove, IL 60515, USA

Whatman International Ltd, Maidstone, Kent, UK

Index